The BEST WRITING on MATHEMATICS

2017

The BEST WRITING on MATHEMATICS

Mircea Pitici, Editor

PRINCETON UNIVERSITY PRESS
PRINCETON AND OXFORD

Published by Princeton University Press, 41 William Street,
Princeton, New Jersey 08540
In the United Kingdom: Princeton University Press, 6
Oxford Street, Woodstock, Oxfordshire OX20 1TW

press.princeton.edu

ISBN (pbk.) 978-0-691-17863-9

British Library Cataloging-in-Publication Data is available

This book has been composed in Perpetua

Printed on acid-free paper. ∞

Printed in the United States of America

1 3 5 7 9 10 8 6 4 2

Dedicated to Barbara Stripling,
for professional inspiration

Contents

Color illustrations follow page 110

Introduction

Mircea Pitici

The eighth volume of *The Best Writing on Mathematics* brings you a new collection of diverse, surprising, and well-written pieces, all published originally during 2016 in academic journals, scientific magazines, or mass media. In addition to the selection, at the end of the book I offer a copious reference section of notable writings and sources for those of you who want to find out more about mathematics on your own; that supplement is important for the goals of the series, serving the research needs of interested readers.

I hope that this series illustrates the versatility of mathematics and that of its interpretations; I also hope that the series helps readers gain a rich panoramic view of mathematics, as opposed to the impoverished parochial view promoted at all levels by our education system. The more facets of mathematics we discover, the more aware we are that mathematics has become a behemoth of human thought, with tentacles reaching into many of the ingenious innovations that fill our personal and collective lives with technological wonders and with deadly perils.

In a memorable line from the movie *Stand and Deliver* (1988), the mathematics teacher Jaïme Escalante tells his students in a run-down Los Angeles community that "math is the great equalizer"—meaning, perhaps, that learning mathematics opens up life possibilities for achievement to everyone, regardless of their ethnicity, social condition, and family status. My own avatars teach me that, like everything else people say about mathematics, Escalante's proclamation is both true *and* untrue, depending on the perspective one takes and on the life caprices one encounters. Against a long and parsimonious tradition that associates mathematics with recipes of ready-made clichés, I aim to show with this series that mathematics is more interesting than the most interesting writing about it—and, more, that this statement remains valid even if we replace the attribute "interesting" with one of its antonyms, or even with most other attributes. Such a sweeping statement will sound disconcerting to the unaware mind, but it is intriguing to the inquisitive mind. Mathematics is a domain of clarity *and* obscurity, of enchantment *and* boredom, of unperturbed neatness *and* of puzzling paradox, of apodictic truth *and* of arguable interpretation. The pieces collected in this volume once again demonstrate the dynamic coexistence of opposite characteristics of mathematics—and show that mathematics is anything but the dull subject serviced by an increasingly powerful but stultifying educational bureaucracy unable to grasp, appreciate, promote, and teach the creative and imaginative sides of mathematics.

Overview of the Volume

In the same vein as the previous books in the series, this volume contains both expository and interpretive pieces on mathematics and aspects of life in the mathematical community, historical and contemporary.

To open the selection, Philip Davis sees mathematicians as producers and shows, with many examples from the past and from the present, that treating mathematical results as "products" is neither far-fetched nor outrageous; it is just an observation supported by abundant evidence but still denied by many mathematicians.

Evelyn Lamb explains why it is useful to know that certain big numbers are primes—and why people are finding primes among a variety of numbers of certain algebraic expression.

Kevin Hartnett describes the work of geometers and physicists who attempt to discover similarities among various random processes.

Siobhan Roberts glosses on the idiosyncratic mathematical achievements of a peculiar centenarian, the recently celebrated Richard Guy.

Lloyd Trefethen shows that the precision of mathematical statements obscures the multitude of contexts in which we can interpret such results.

Gerald Alexanderson reviews biographical contributions inspired by Srinivasa Ramanujan's life and work, and he tells us the intriguing circumstances under which he acquired a bronze bust of the famous Indian mathematician.

Larry Riddle brings abstract algebra to the study of systems of functions to create beautiful fractal images.

Marc Frantz contributes with elements of projective geometry to the wider context (within optics and perception) of the moon tilt illusion.

Mohammadhossein Kasraei, Yahya Nourian, and Mohammadjavad Mahdavinejad study how the Persian architectural element Girih was used in the construction of three Iranian domes; they also analyze the relationship between dome curvature and the polygonal division of the dome's base circle.

Jo Boaler and Lang Chen summarize studies from several disciplines to conclude that children's degree of dexterity with "finger math" is important to their mathematical development.

Sinéad Breen and Ann O'Shea rethink the design of undergraduate mathematics education, proposing that the central role held by the pairing of content and techniques should be replaced by "threshold concepts," which they define and characterize in their piece.

John Mason pleads for a mathematics education attentive to the circumstantial elements that occasion learning—as opposed to the dogmatism of normative theories so popular with researchers.

Viktor Blåsjö exemplifies with a geometric-algebraic construction taken from Leibniz's work the changing meaning of mathematics and mathematics notation over the past few centuries.

Carlo Séquin and Raymond Shiau examine a famous painting by Fra' Luca Pacioli to determine whether the plane rendering of a spatial geometric object is genuine, and they bring the topic to the present by offering a computerized version of that representation.

Jeremy Gray asks what would have passed as most valuable mathematical research, most worthy of award-winning consideration, a century and a half ago and examines in that context the work of several mathematicians prominent at the time.

Noson Yanofsky illustrates with an abundance of examples different types of mathematical and scientific limitations, from logical and physical to mental and practical.

Jean-Pierre Marquis defines abstraction and "levels" of abstraction in mathematics, distinguishing between the axiomatic method and the abstract method. Then he infers the philosophical consequences of using the latter.

Robert Bain considers pro and con arguments for the proposition that human reasoning, beliefs, and decision making actively adjust based on evidence and probability expectations.

Speaking of expectations, in the last piece of the anthology, Graham Southorn describes quantitative methods used in forecasting and explains why they never achieve certainty in our ever more complex world.

More Writings on Mathematics

Among recent books on mathematics deserving special mention are the following: the *Sourcebook in the Mathematics of Medieval Europe and North Africa* edited by Victor Katz; the path-opening *Visualizing Mathematics with 3D Printing* by Henry Segerman; the book-and-catalogue *Mathematics* edited by David Rooney for the Science Museum of London; and two interdisciplinary books reaching both close to and far away from mathematics, *The Oxford Handbook of Generality in Mathematics and the Sciences* edited by Karine Chemla, Renaud Chorlay, and David Rabouin, and the massive *Handbook of Geomathematics* edited by Willi Freeden, Zuhair Nashed, and Thomas Sonar.

Expository books on mathematics are *Elements of Mathematics* by John Stillwell, *The Circle* by Alfred Posamentier and Robert Geretschläger, *Algebra* by Peter Higgins, *Fractals* by Kenneth Falconer, *Combinatorics* by Robin Wilson, *Measurement* by David Hand, *Some Applications of Geometric Thinking* by Bowen Kerins et al., *Thinking Geometrically* by Thomas Sibley, *Geometry in Problems* by Alexander Shen, *Problem-Solving Strategies in Mathematics* by Alfred Posamentier and Stephen Krulik, *An Interactive Introduction to Knot Theory* by Inga Johnson and Allison Henrich, *Circularity* by Ron Aharoni, *Can You Solve My Problems?* by Alex Bellos, and *Summing It Up* by Avner Ash and Robert Gross.

Mathematics in life (including gambling and games) is described and interpreted in such books as *The Calculus of Happiness* by Oscar Fernandez, *Fluke* by Joseph Mazur, *Man vs. Mathematics* by Timothy Revell and Joe Lyward, *In Praise of Simple Physics* by Paul Nahin, *The Mathematics that Power Our World* by Joseph Khoury and Gilles Lamothe, *Living by Numbers* by Steven Connor, *The Perfect Bet* by Adam Kucharski, *The Joy of SET* by Liz McMahon and her coauthors, *That's Maths* by Peter Lynch, and *Math Squared* by Rachel Thomas and Maryanne Freiberger; Daniel Levitin takes a broad perspective in *A Field Guide to Lies*.

In the history of mathematics and biography, recently I noticed *A Brief History of Mathematical Thought* by Luke Heaton, *Infinite Series in a History of Analysis* by Hans-Heinrich Körle, *Turing* by Jack Copeland,

What Is the Genus? by Patrick Popescu-Pampu, *A Delicate Balance* edited by David Rowe and Wann-Sheng Horng, *The Early Period of the Calculus of Variations* by Paolo Freguglia and Mariano Giaquinta and a new edition of *Emmy Noether's Wonderful Theorem* by Dwight Neuenschwander.

Several recent books on mathematical connections with other disciplines are Dmitry Kondrashov's *Quantitative Life*, Richard Harris's *Quantitative Geography*, Youseop Shin's *Time Series Analysis in the Social Sciences*, and Ron Aharoni's *Mathematics, Poetry, and Beauty*. Collective volumes include *Big Data and Social Science* edited by Ian Foster and his collaborators, *Big Data in Cognitive Science* edited by Michael Jones, *Data Visualization* edited by Lauren Magnuson, *Handbook of Quantitative Methods for Detecting Cheating on Tests* edited by Gregory Cizek and James Wollack, *UK Success Stories in Industrial Mathematics* edited by Philip Aston, Anthony Mulholland and Katherine Tant, and *From Numbers to Words* by Susan Morgan, Tom Reichert, and Tyler Harrison. Two interdisciplinary volumes are *Quite Right* by Norman Biggs and *The Topological Imagination* by Angus Fletcher.

Many more books than I can mention are published each year in mathematics education. Some recent volumes that came to my attention are *Creativity and Giftedness* edited by Roza Leikin and Bharath Sriraman, *Posing and Solving Mathematical Problems* edited by Patricio Felmer, Erkki Pehkonen, and Jeremy Kilpatrick, *Teaching School Mathematics* by Hung- Hsi Wu, *Task Design in Mathematics Education* edited by Anne Watson and Minoru Ohtani, *Semiotics as a Tool for Learning Mathematics* edited by Adalira Sáenz-Ludlow and Gert Kadunz, *Psychometric Methods in Mathematics Education* edited by Andrew Izsák, Janine Remillard, and Jonathan Templin, *Mathematics Education and Language Diversity* edited by Richard Barwell et al., *Putting Essential Understanding of Geometry and Measurement into Practice* edited by Kathryn Chval, and *The Second Handbook of Research on the Psychology of Mathematics Education* edited by Ángel Gutiérrez, Gilah C. Leder, and Paolo Boero.

To conclude, I enumerate several recent titles in the philosophy of mathematics and some logic essays: *Making and Breaking Mathematical Sense* by Roi Wagner, *Essays on Paradoxes* by Terence Horgan, *Bolzano's Logical System* by Ettore Casari, *Talking about Numbers* by Katharina Felka, *Resonance* by Krzysztof Burdzy, *Plural Logic* by Alex Oliver and Timothy Smiley, a new volume in Yuki Hiroshi's *Math Girls* series, as well as the collective volumes *Logical Modalities from Aristotle to Carnap* edited by Max

Cresswell, Edwin Mares, and Adriane Rini, and *Cultures of Mathematics and Logic* edited by Shier Ju, Benedikt Löwe, and their collaborators. A sociological viewpoint underpins *The Quantified Self* by Deborah Lupton.

I hope that you, the reader, will enjoy reading this anthology at least as much as I did while working on it. I encourage you to send comments, suggestions, and materials I might consider for (or mention in) future volumes to Mircea Pitici, P.O. Box 4671, Ithaca, NY 14852; or send electronic correspondence to mip7@cornell.edu.

Books Mentioned

Aharoni, Ron. *Circularity: A Common Secret to Paradoxes, Scientific Revolutions, and Humor.* World Scientific, Singapore: 2016.

Aharoni, Ron. *Mathematics, Poetry, and Beauty.* Singapore: World Scientific, 2016.

Ash, Avner, and Robert Gross. *Summing It Up: From One Plus One to Modern Number Theory.* Princeton, NJ: Princeton University Press, 2016.

Aston, Philip J, Anthony J. Mulholland and Katherine M. M. Tant. (Eds.) *UK Success Stories in Industrial Mathematics.* New York: Springer Science+Business Media, 2016

Barwell, Richard, et al. (Eds.) *Mathematics Education and Language Diversity.* Cham, Switzerland: Springer Science+Business Media, 2016.

Bellos, Alex. *Can You Solve My Problems? A Casebook of Ingenious, Perplexing, and Totally Satisfying Puzzles.* London, UK: Guardian Books, 2016.

Biggs, Norman. *Quite Right: The Story of Mathematics, Measurement, and Money.* Oxford, UK: Oxford University Press, 2016.

Burdzy, Krzysztof. *Resonance: From Probability to Epistemology and Back.* London: Imperial College Press, 2016.

Casari, Ettore. *Bolzano's Logical System.* Oxford, UK: Oxford University Press, 2016.

Chemla, Karine, Renaud Chorlay, and David Rabouin. (Eds.) *The Oxford Handbook of Generality in Mathematics and the Sciences.* Oxford, UK: Oxford University Press, 2016.

Chval, Kathryn. (Ed.) *Putting Essential Understanding of Geometry and Measurement into Practice in Grades 3–5.* Reston, VA: The National Council of Teachers of Mathematics, 2016.

Cizek, Gregory J., and James A. Wollack. (Eds.) *Handbook of Quantitative Methods for Detecting Cheating on Tests.* Abingdon, UK: Routledge, 2016.

Connor, Steven. *Living by Numbers: In Defense of Quantity.* London: Reaktion Books, 2016.

Copeland, Jack. *Turing: Pioneer of the Information Age.* Oxford, UK: Oxford University Press, 2014.

Cresswell, Max, Edwin Mares, and Adriane Rini. (Eds.) *Logical Modalities from Aristotle to Carnap: The Story of Necessity.* Cambridge, UK: Cambridge University Press, 2016.

Falconer, Kenneth. *Fractals: A Very Short Introduction.* Oxford, UK: Oxford University Press, 2013.

Felka, Katharina. *Talking about Numbers: Easy Arguments for Mathematical Realism.* Frankfurt, Germany: Klostermann, 2016.

Felmer, Patricio, Erkki Pehkonen, and Jeremy Kilpatrick. (Eds.) *Posing and Solving Mathematical Problems: Advances and New Perspectives.* Cham, Switzerland: Springer Nature, 2016.

Fernandez, Oscar E. *The Calculus of Happiness: How a Mathematical Approach to Life Adds Up to Health, Wealth, and Love.* Princeton, NJ: Princeton University Press, 2017.

Fletcher, Angus. *The Topological Imagination: Spheres, Edges, and Islands.* Cambridge, MA: Harvard University Press, 2016.

Foster, Ian, et al. (Eds.) *Big Data and Social Science: A Practical Guide to Methods and Tools.* Boca Raton, FL: CRC Press, 2016.

Freeden, Willi, M. Zuhair Nashed, and Thomas Sonar. (Eds.) *Handbook of Geomathematics.* Heidelberg, Germany: Springer Verlag, 2015.

Freguglia, Paolo, and Mariano Giaquinta. *The Early Period of the Calculus of Variations.* Cham, Switzerland: Springer Science+Business Media, 2016.

Gutiérrez, Ángel, Gilah C. Leder, and Paolo Boero. (Eds.) *The Second Handbook of Research on the Psychology of Mathematics Education: The Journey Continues.* Rotterdam, Netherlands: Sense Publishers, 2016.

Hand, David J. *Measurement: A Very Short Introduction.* Oxford, UK: Oxford University Press, 2016.

Harris, Richard. *Quantitative Geography: The Basics.* Los Angeles: Sage, 2016.

Heaton, Luke. *A Brief History of Mathematical Thought.* Oxford, UK: Oxford University Press, 2017.

Higgins, Peter M. *Algebra: A Very Short Introduction.* Oxford, UK: Oxford University Press, 2015.

Hiroshi, Yuki. *Math Girls: Gödel's Incompleteness Theorem.* Austin, TX: Benton Books, 2016.

Horgan, Terence. *Essays on Paradoxes.* Oxford, UK: Oxford University Press, 2017.

Izsák, Andrew, Janine T. Remillard, and Jonathan Templin. (Eds.) *Psychometric Methods in Mathematics Education: Opportunities, Challenges, and Interdisciplinary Collaborations.* Reston, VA: The National Council of Teachers of Mathematics, 2016.

Johnson, Inga, and Allison Henrich. *An Interactive Introduction to Knot Theory.* Mineola, NY: Dover Publications, 2016.

Jones, Michael N. (Ed.) *Big Data in Cognitive Science.* New York: Routledge, 2017.

Katz, Victor J. (Ed.) *Sourcebook in the Mathematics of Medieval Europe and North Africa.* Princeton, NJ: Princeton University Press, 2016.

Kerins, Bowen, et al. *Some Applications of Geometric Thinking.* Providence, RI: American Mathematical Society, 2016.

Khoury, Joseph and Gilles Lamothe. *The Mathematics that Power Our World: How Is It Made?* Singapore: World Scientific, 2016.

Kondrashov, Dmitry A. *Quantitative Life: A Symbiosis of Computation, Mathematics, and Biology.* Chicago: Chicago University Press, 2016.

Körle, Hans-Heinrich. *Infinite Series in a History of Analysis: Stages Up to the Verge of Summability.* Berlin: de Gruyter, 2015.

Kucharski, Adam. *The Perfect Bet: How Science and Math Are Taking the Luck Out of Gambling.* New York: Basic Books, 2016.

Leikin, Roza, and Bharath Sriraman. (Eds.) *Creativity and Giftedness: Interdisciplinary Perspectives from Mathematics and Beyond.* Basel, Switzerland: Springer, 2016.

Levitin, Daniel J. *A Field Guide to Lies: Critical Thinking in the Information Age.* New York: Dutton, 2016.

Lupton, Deborah. *The Quantified Self: A Sociology of Self-Tracking.* Cambridge, UK: Polity Press, 2016.

Lynch, Peter. *That's Maths.* Dublin, Ireland: M. H. Gill & Co., 2016.

Magnuson, Lauren. (Ed.) *Data Visualization: A Guide to Visual Storytelling for Libraries.* Lanham, MD: Rowman & Littlefield, 2016.

Mazur, Joseph. *Fluke: The Math and Myth of Coincidence*. New York: Basic Books, 2016.
McMahon, Liz, Gary Gordon, Hannah Gordon, and Rebecca Gordon. *The Joy of SET: The Many Mathematical Dimensions of a Seemingly Simple Card Game*. Princeton, NJ: Princeton University Press, 2017.
Morgan, Susan E., Tom Reichert, and Tyler R. Harrison. *From Numbers to Words: Reporting Statistical Results for the Social Sciences*. New York: Routledge, 2017.
Nahin, Paul J. *In Praise of Simple Physics: The Science and Mathematics behind Everyday Questions*. Princeton, NJ: Princeton University Press, 2017.
Neuenschwander, Dwight E. *Emmy Noether's Wonderful Theorem*. Baltimore: The Johns Hopkins University Press, 2017.
Oliver, Alex, and Timothy Smiley. *Plural Logic*, 2nd ed. Oxford, UK: Oxford University Press, 2016.
Popescu-Pampu, Patrick. *What Is the Genus?* Cham, Switzerland: Springer Nature, 2016.
Posamentier, Alfred S., and Robert Geretschläger. *The Circle: A Mathematical Exploration beyond the Line*. Amherst, NY: Prometheus Books, 2016.
Posamentier, Alfred S., and Stephen Krulik. *Problem-Solving Strategies in Mathematics: From Common Approaches to Exemplary Strategies*. Singapore: World Scientific, 2015.
Revell, Timothy, and Joe Lyward. *Man vs. Mathematics: Understanding the Curious Mathematics That Powers Our World*. London: Aurum Press, 2016.
Rooney, David. (Ed.) *Mathematics: How It Shaped Our World*. London: Scala Arts & Heritage Publishers and The Science Museum, 2016.
Rowe, David, and Wann-Sheng Horng. (Eds.) *A Delicate Balance: Global Perspectives on Innovation and Tradition in the History of Mathematics*. Bern, Switzerland: Birkhäuser, 2016.
Sáenz-Ludlow, Adalira, and Gert Kadunz. (Eds.) *Semiotics as a Tool for Learning Mathematics: How to Describe the Construction, Visualization, and Communication of Mathematical Concepts*. Rotterdam, Netherlands: Sense Publishers, 2016.
Segerman, Henry. *Visualizing Mathematics with 3D Printing*. Baltimore: The Johns Hopkins University Press, 2016.
Shen, Alexander. *Geometry in Problems*. Providence, RI: American Mathematical Society, 2016.
Shier Ju, Benedikt Löwe, Thomas Müller, and Yun Xie. (Eds.) *Cultures of Mathematics and Logic*. Bern, Switzerland: Birkhäuser, 2016.
Shin, Youseop. *Time Series Analysis in the Social Sciences: The Fundamentals*. Oakland, CA: University of California Press, 2017.
Sibley, Thomas Q. *Thinking Geometrically: A Survey of Geometries*. Washington, DC: Mathematical Association of America, 2015.
Stillwell, John. *Elements of Mathematics: From Euclid to Gödel*. Princeton, NJ: Princeton University Press, 2016.
Thomas, Rachel, and Maryanne Freiberger. *Math Squared: 100 Ideas You Should Know*. London: Metro, 2016.
Wagner, Roi. *Making and Breaking Mathematical Sense: Histories and Philosophies of Mathematical Practice*. Princeton, NJ: Princeton University Press, 2016.
Watson, Anne, and Minoru Ohtani. (Eds.) *Task Design in Mathematics Education*. Cham, Switzerland: Springer Science+Business Media, 2016.
Wilson, Robin. *Combinatorics: A Very Short Introduction*. Oxford, UK: Oxford University Press, 2016.
Wu, Hung-Hsi. *Teaching School Mathematics: Pre-Algebra*. Providence, RI: American Mathematical Society, 2016.

The **BEST** **WRITING** on **MATHEMATICS**

2017

Mathematical Products

Philip J. Davis

A prominent mathematician recently sent me an article he had written and asked me for my reaction. After studying it, I said that he was proposing a mathematical "product" and that as such it stood in the scientific marketplace in competition with nearby products. He bridled and was incensed by my use of the word "product" to describe his work. Our correspondence terminated. What follows is an elaboration of what I mean by mathematical products and how I situate them within the mathematical enterprise.

Civilization has always had a mathematical underlay, often informal, and not always overt. I would say that mathematics often lies deep in formulaic material, procedures, conceptualizations, attitudes, and now in chips and accompanying hardware. In recent years, the mathematization of our lives has grown by leaps and bounds. A useful point of view is to think of this growth in terms of products. Mathematical products serve a purpose; they can be targeted to define, facilitate, enhance, supply, explain, interpret, invade, complicate, confuse, and create new requirements or environments for life.

What? Mathematical "products"? Products in an intellectual area that is reputed to contain the finest result of pure reason and logic: a body of material that in its early years was in the classical quadrivium along with astronomy and music? How gross of me to bring in the language of materialistic commerce and in this way sully or besmirch the reputation of what are clean, crisp idealistic constructions! Products are the routine output of factories, not of skilled craft workers whose sharp minds frequently reside far above the usual rewards of life. The notion that mathematics has products or that its content is merchandise might tarnish both its image and the self-image of the creators of this noble material.

And yet . . . the world is full of mathematical products—mathematics *produces* knowledge, hence we have *mathematical products*—many of them. As O'Halloran [1, 2] claims, mathematics is functional; it permits us to construe and reason about the world in new ways that extend beyond our linguistic formulations. The world of today embraces the product of that knowledge.

1 Examples of Mathematical Products

Yes, the world is full of mathematical products of all sorts. I will name a few. A slide rule is a product. A French curve is a product. An algorithm (recipe) for solving linear equations is a product. A theorem is a product and stands among hundreds of thousands of theorems, ready to be interpreted, appreciated, used, updated, reworked, or neglected. A textbook on linear algebra is a product. A polling system is a product. The statutory rule for allocating representatives after a new census is a product. A tax or a lottery or an insurance scheme, or even a Ponzi scheme is a product. Telephone numbers are products. A professional mathematical society is a product. A medical decision that depends in a routine manner on some sort of quantification is a product. A computer language is a product. A supermarket cash register and the Julian calendar are products. The act of taking a number at a delicatessen or a bakeshop counter to facilitate one's "next" is a product. MATLAB is a product. Google is a product. Encryption schemes are products. Sometimes a mathematical product is designed for very specialized usage; it may then be called a package or a toolbox.

Admittedly, these examples might seem to indicate that in my mind anything at all that has to do with mathematics can be considered a product. Is Cantor's diagonalization process a product? Is a T-shirt imprinted with the face of Kurt Gödel a mathematical product? Well, I would find it exceedingly difficult to propose a formal definition. In any case, let us see the extent to which one might describe and discuss the mathematical enterprise from the point of view of its products that I have cited or will cite.

What is the clientele for mathematical products? While mathematical products are the brain children of inspired individuals or groups, the targeted users of the products may vary from a few individuals to entire populations. Those targeted may be aware of the availability of a product that has been claimed to be of use; they may either use it or reject it. In many cases, the product is built into a whole social system and one cannot easily opt out of its use. Examples include phone numbers and area codes, the U.S. Census, and more locally, passwords at the ATM around the corner.

1.1 Scientific and Technological Aspects of Mathematical Products

Mathematics was called by Gauss "the Queen of the Sciences," and a good fraction of its products relate to science and technology, e.g., packages for the factorization of large integers; for the analysis of architectural structures or packages marketed; or for on-site DNA analysis. A scheme for constructing and interpreting a horoscope can be a mathematical product of considerable sophistication and complexity. The "wise" may reject its conclusions, yet the product flourishes.

Without in any way dethroning the Queen, it should be pointed out that the employment of mathematics has always gone far beyond what are now called "the sciences." Mathematics has had an effect on commerce, trade, medicine, biology, mysticism, theology, entertainment, and warfare.

1.2 The Transmission or Communication of Mathematical Products

Transmission is done by a wide variety of "signs" or "semiotic products." Short texts, books, pictures, programs, flash drives, chips, formal classroom teaching, the informal master–apprentice relationship, and word of mouth. The international or intercultural transmission and absorption of mathematical products (e.g., the adoption by the West of Arabic numerals) has been and still is the object of scholarly studies.

1.3 Commercial Aspects of Mathematical Products

The commercialization of mathematical products has grown by leaps and bounds since World War II. A mathematical product can be promoted in many of the same ways that a brand of breakfast food is promoted: by ads or by the praises of well-known personalities or groups (plugs). On MATLAB's website, you can find a list of MATLAB's available products, listed openly and labeled clearly as "products." Investment and insurance schemes are called "products."

A product can be sold, e.g., a handheld computer or the *Handbook of Mathematical Functions*. A product can be licensed for usage, or it can be made available as a freebie. In the case of taxes (*qua* mathematical product), it is "promoted" by laws and threats of punishment. Rubik's Cube, a mathematical product, caught the imagination and challenged the wits of millions of people and has earned fortunes. Sudoku, a mathematical puzzle, is sold in numerous formats. If a product is income producing, its sellers can be taxed. A product can be copyrighted or patented; the owners of such can be contested or sued for infringement.

1.4 Competitive Aspects of Mathematical Products

A mathematical product is often subject to competition from nearby products. Think of the innumerable ways of solving a set of linear equations. Textbooks, a source of considerable income, compete in a mathematical marketplace that involves educationists, testing theorists and outfits, unions, publishers, parents' groups, and local state and national governments.

1.5 Social Aspects of Mathematical Products

If a mathematical product finds widespread usage, it may have social, economic, ethical, legal, or political implications or consequences. The repugnant Nuremberg Racial Laws in Germany in 1935, with their numerical criteria, caused incredible suffering. DNA sequencing and its interpretations is a relatively new branch of applied mathematics, resulting in a host of new products. In a number of states, the level of mathematical tests for the lower school grades has been questioned. The social consequences of mathematical products, benign or otherwise, may not emerge for many years.

1.6 Legal Aspects of Mathematical Products

There are innumerable examples of this. The U.S. Constitution is full of number processes. Consider

> Representatives and direct Taxes shall be apportioned among the several States which may be included within this Union, according to their respective Numbers, which shall be determined by adding to the whole Number of free Persons, including those bound to Service for a Term of Years, and excluding Indians not taxed, three fifths of all other Persons. (Later Amended!)

Some mathematical products have been subject to judicial review. As an example, the mathematical scheme for the 2010 Census was vetted and restricted by the U.S. Supreme Court.

An example of a statutory product is the method of least proportions used to allocate representatives in Congress. It was approved by the Supreme Court in *Department of Commerce v. Montana*, 503 U.S. 442 (1992). A multiple regression model used in an employment discrimination class action is another such example; it was approved by the Supreme Court in *Bazemore v. Friday*, 478 U.S. 385 (1986).

1.7 Logical or Philosophical Aspects of Mathematical Products

A mathematical product, considered as such, is neither true nor false. Of course, it may embody certain principles of deductive logic, but these do not automatically make the employment of the product plausible or advisable. A product can be made plausible, moot, or useless on the basis of certain internal or external considerations. An interesting historical example of this is the dethroning of Euclidean geometry as the unique geometry by the discovery of non-Euclidean geometries.

A product may raise or imply philosophical questions, such as the distinction between the subjective and the objective or between the qualitative and the quantitative, between the deterministic and the probabilistic, the tangible and the intangible, the hidden and the overt.

Numerical indexes of this thing and that thing abound. Cases of subjectivity occur when a product asks a person or a group of people to pass judgment on some issue: "On a scale of zero to ten, how much do you like tofu?" The well-known *Index of Economic Freedom* embodies a number of items, expressed numerically:

> We measure ten components of economic freedom, assigning a grade in each using a scale from 0 to 100, where 100 represents

the maximum freedom. The ten component scores are then averaged to give an overall economic freedom score for each country. The ten components of economic freedom are: Business Freedom | Trade Freedom | Fiscal Freedom | Government Size | Monetary Freedom | Investment Freedom | Financial Freedom | Property Rights | Freedom from Corruption | Labor Freedom

1.8 Moral Aspects of Mathematical Products

Society asks many questions. Does the manner of taking the U.S. Census account properly for the homeless? Are tests in algebra slanted toward certain subcultures? Does the tremendous role that mathematics plays in war raise questions or angst in the minds of those who are responsible for its application? Are results of IQ testing being misused?

2 Judgments of Mathematical Products

As mentioned, mathematical products serve a purpose; they can be targeted to define, facilitate, enhance, or invade any of the requirements or aspects of life. Ultimately, a mathematical product can be judged in the same way that any product can be judged: by the response of its targeted users or purchasers. In the case of a mathematical product, what criteria are in play? The cheapest? The most convenient? The most useful? The most comprehensive? The most accurate? The most original? The most seminal? The most reassuring? The safest or least vulnerable? The most esthetic? The most moral? Is the product unique? Are there pressures from investors or the various foundations that support their production?

Is "survival of the fittest" a good description of the judgment process? Probably not. There are fashions in the product world attracting both excited consumers and producers. Time, chance, and what the larger world requires, appreciates, or suffers from mathematizations are always in play to determine survival.

References

1. O'Halloran, K. L. (2005). *Mathematical Discourse: Language, Symbolism and Visual Images*. London and New York: Continuum.
2. O'Halloran, K. L. (2015). "The Language of Learning Mathematics: A Multimodal Perspective." *The Journal of Mathematical Behaviour*, http://clx.doi.org/10.1016/j.jmathb.2014.09.002.

The Largest Known Prime Number

Evelyn Lamb

Earlier this week, BBC News reported an important mathematical finding, sharing the news under the headline "Largest Known Prime Number Discovered in Missouri." That phrasing makes it sound a bit like this new prime number—it's $2^{74,207,281} - 1$, by the way—was found in the middle of some road, underneath a street lamp. That's actually not a bad way to think about it.

We know about this enormous prime number thanks to the Great Internet Mersenne Prime Search. The Mersenne hunt, known as GIMPS, is a large distributed computing project in which volunteers run software to search for prime numbers. Perhaps the best-known analogue is SETI@home, which searches for signs of extraterrestrial life. GIMPS has had a bit more tangible success than SETI, with 15 primes discovered so far. The shiny new prime, charmingly nicknamed M74207281, is the fourth found by University of Central Missouri mathematician Curtis Cooper using GIMPS software. This one is 22,338,618 digits long.

A prime number is a whole number whose only factors are 1 and itself. The numbers 2, 3, 5, and 7 are prime, but 4 is not because it can be factored as 2×2. (For reasons of convenience, we don't consider 1 to be a prime.) The M in GIMPS and in M74207281 stands for Marin Mersenne, a 17th-century French friar who studied the numbers that bear his name. Mersenne numbers are 1 less than a power of 2. Mersenne primes, logically enough, are Mersenne numbers that are also prime. The number 3 is a Mersenne prime because it's one less than 2^2, which is 4. The next few Mersenne primes are 7, 31, and 127.

M74207281 is the 49th known Mersenne prime. The next largest known prime, $2^{57,885,161} - 1$, is also a Mersenne prime. So is the one after that. And the next one. And the next one. All in all, the 11 largest known primes are Mersenne.

Why do we know about so many large Mersenne primes and so few large non-Mersenne ones? It's not because large Mersenne primes are particularly common, and it's not a spectacular coincidence. That brings us back to the road and the street lamp. There are several different versions of the story. A man, perhaps he's drunk, is on his hands and knees underneath a streetlight. A kind passerby, perhaps a police officer, stops to ask what he's doing. "I'm looking for my keys," the man replies. "Did you lose them over here?" the officer asks. "No, I lost them down the street," the man says, "but the light is better here."

We keep finding large Mersenne primes because the light is better there.

First, we know that only a few Mersenne numbers are even candidates for being Mersenne primes. The exponent n in $2^n - 1$ needs to be prime, so we don't need to bother to check $2^6 - 1$, for example. There are a few other technical conditions that make certain prime exponents more enticing to try. Finally, there's a particular test of primeness—the Lucas–Lehmer test—that can only be used on Mersenne numbers.

To understand why the test even exists, let's take a detour to explore why we bother finding primes in the first place. There are infinitely many of them, so it's not like we're going to eventually find the biggest one. But aside from being interesting in a "math for math's sake" kind of way, finding primes is good business. RSA encryption, one of the standard ways to scramble data online, requires the user (perhaps your bank or Amazon) to come up with two big primes and multiply them together. Assuming that the encryption is implemented correctly, the difficulty of factoring the resulting product is the only thing between hackers and your credit card number.

This new Mersenne prime is not going to be used for encryption any time soon. Currently, we only need primes that are a few hundred digits long to keep our secrets safe, so the millions of digits in M74207281 are overkill, for now.

You can't just look up a 300-digit prime in a table. (There are about 10^{297} of them. Even if we wanted to, we physically could not write them all down.) To find large primes to use in RSA encryption, we need to test randomly generated numbers for primality. One way to do this is trial division: Divide the number by smaller numbers and see if you ever get a whole number back. For large primes, this takes way too long. Hence, there are primality tests that can determine whether a number is prime without actually factoring it. The Lucas–Lehmer test is one of the best.

The heat death of the universe would occur before we could get even a fraction of the way through trial division of a number with 22 million digits. It only took a month, however, for a computer to use the Lucas–Lehmer test to determine that M74207281 is prime. There are no other primality tests that run nearly as quickly for arbitrary 22 million–digit numbers.

How many primes have we missed by looking for them mostly under the Lucas–Lehmer street lamp? We don't know the exact answer, but the prime number theorem gets us close enough. It makes sense that primes become less common as we stroll out on the number line. Fully 40% of one-digit numbers are prime, 22% of two-digit numbers are prime, and only 16% of three-digit numbers are. The prime number theorem, first proved in the late 1800s, quantifies that decline. It says that in general, the number of primes less than n tends to $n/\ln(n)$ as n increases. (Here ln is the natural logarithm.)

We can use the prime number theorem to estimate how many missing primes there are between M74207281 and the next smallest known prime. We just plug $2^{74,207,281} - 1$ into $n/\ln(n)$ and get, well, a really big number. We can write it most compactly by stacking exponents: $10^{10^{7.349}}$. This number has about 22,338,610 digits, give or take a couple, so we can also write it as $10^{22,338,610}$.

Another visit to the prime number theorem shows that there are approximately $10^{17,425,163}$ primes less than the next-largest known prime. That sounds impressive until you realize that $10^{17,425,163}$ is less than 0.000000000001% of $10^{22,338,610}$.

Stop and think about that for a moment. There are about $10^{22,338,610}$ primes less than M74207281, and approximately all of them are between it and the next-largest known prime. If you want to be charitable, you could say that we have some gaps in our knowledge of prime numbers. But really, it makes more sense to say that we have gaps in our lack of knowledge. The millions upon millions of prime numbers we've already found make up approximately 0% of the primes that are less than M74207281. Each one is a little grain of sand, a speck that does little to cover up our overwhelming ignorance of exactly where the prime numbers live.

A Unified Theory of Randomness

Kevin Hartnett

Standard geometric objects can be described by simple rules—every straight line, for example, is just $y = ax + b$—and they stand in neat relation to each other: Connect two points to make a line, connect four line segments to make a square, connect six squares to make a cube.

These are not the kinds of objects that concern Scott Sheffield. Sheffield, a professor of mathematics at the Massachusetts Institute of Technology, studies shapes that are constructed by random processes (Figure 1). No two of them are ever exactly alike. Consider the most familiar random shape, the random walk, which shows up everywhere from the movement of financial asset prices to the path of particles in quantum physics. These walks are described as random because no knowledge of the path up to a given point can allow you to predict where it will go next.

Beyond the one-dimensional random walk, there are many other kinds of random shapes. There are varieties of random paths, random two-dimensional surfaces, random growth models that approximate, for example, the way a lichen spreads on a rock. All of these shapes emerge naturally in the physical world, yet until recently they've existed beyond the boundaries of rigorous mathematical thought. Given a large collection of random paths or random two-dimensional shapes, mathematicians would have been at a loss to say much about what these random objects shared in common.

Yet in work over the past few years, Sheffield and his frequent collaborator, Jason Miller, a professor at the University of Cambridge, have shown that these random shapes can be categorized into various classes, that these classes have distinct properties of their own, and that some kinds of random objects have surprisingly clear connections with other kinds of random objects. Their work forms the beginning of a unified theory of geometric randomness.

FIGURE 1. Randomness increases in a structure known as an "SLE curve." Photo by Jason Miller.

"You take the most natural objects—trees, paths, surfaces—and you show they're all related to each other," Sheffield said. "And once you have these relationships, you can prove all sorts of new theorems you couldn't prove before."

In the coming months, Sheffield and Miller will publish the final part of a three-paper series that for the first time provides a comprehensive view of random two-dimensional surfaces—an achievement not unlike the Euclidean mapping of the plane.

"Scott and Jason have been able to implement natural ideas and not be rolled over by technical details," said Wendelin Werner, a professor at ETH Zurich and winner of the Fields Medal in 2006 for his work in probability theory and statistical physics. "They have been basically able to push for results that looked out of reach using other approaches."

A Random Walk on a Quantum String

In standard Euclidean geometry, objects of interest include lines, rays, and smooth curves like circles and parabolas. The coordinate values of the points in these shapes follow clear, ordered patterns that can be described by functions. If you know the value of two points on a line, for instance, you know the values of all other points on the line. The same is true for the values of the points on each of the rays in Figure 2, which begin at a point and radiate outward.

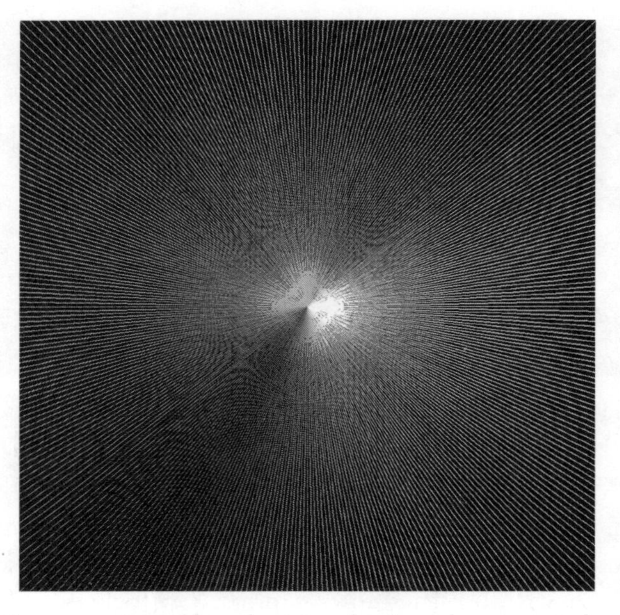

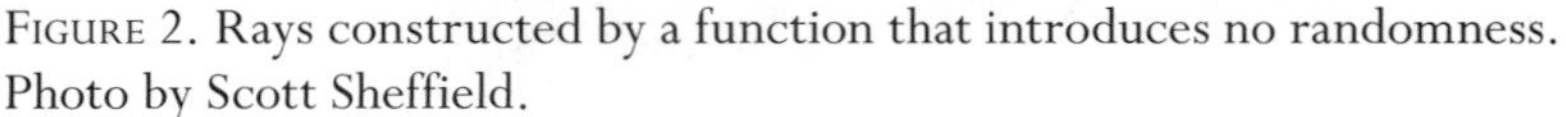

FIGURE 2. Rays constructed by a function that introduces no randomness. Photo by Scott Sheffield.

One way to begin to picture what random two-dimensional geometries look like is to think about airplanes. When an airplane flies a long-distance route, like the route from Tokyo to New York, the pilot flies in a straight line from one city to the other. Yet if you plot the route on a map, the line appears to be curved. The curve is a consequence of mapping a straight line on a sphere (Earth) onto a flat piece of paper.

If Earth were not round, but were instead a more complicated shape, possibly curved in wild and random ways, then an airplane's trajectory (as shown on a flat two-dimensional map) would appear even more irregular, like the rays in Figure 3.

Each ray represents the trajectory an airplane would take if it started from the origin and tried to fly as straight as possible over a randomly fluctuating geometric surface. The amount of randomness that characterizes the surface is dialed up in Figures 4 and 5—as the randomness increases, the straight rays wobble and distort, turn into increasingly jagged bolts of lightning, and become nearly incoherent.

Yet incoherent is not the same as incomprehensible. In a random geometry, if you know the location of some points, you can (at best)

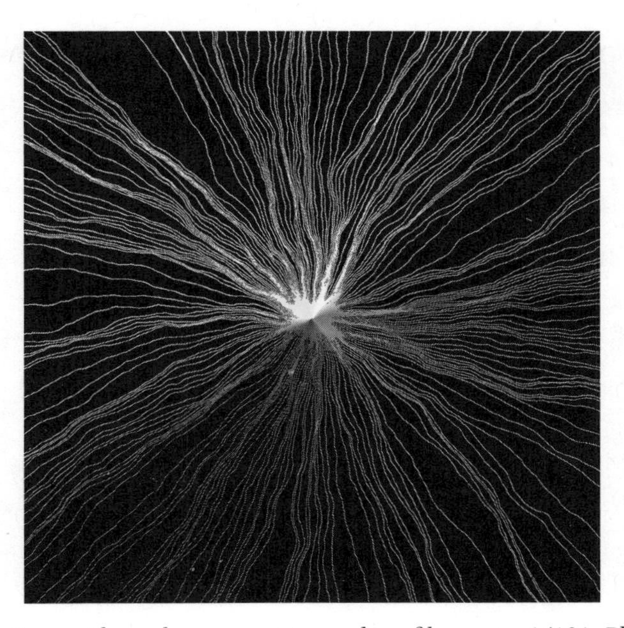

FIGURE 3. Rays with randomness set to a value of kappa = 4/101. Photo by Scott Sheffield. See also color image.

assign probabilities to the location of subsequent points. And just as a loaded set of dice is still random, but random in a different way than a fair set of dice, it's possible to have different probability measures for generating the coordinate values of points on random surfaces.

What mathematicians have found—and hope to continue to find—is that certain probability measures on random geometries are special and tend to arise in many different contexts. It is as though nature has an inclination to generate its random surfaces using a very particular kind of die (one with an uncountably infinite number of sides). Mathematicians like Sheffield and Miller work to understand the properties of these dice (and the "typical" properties of the shapes they produce) just as precisely as mathematicians understand the ordinary sphere.

The first kind of random shape to be understood in this way was the random walk. Conceptually, a one-dimensional random walk is the kind of path you'd get if you repeatedly flipped a coin and walked one way for heads and the other way for tails. In the real world, this type of movement first came to attention in 1827 when the English botanist

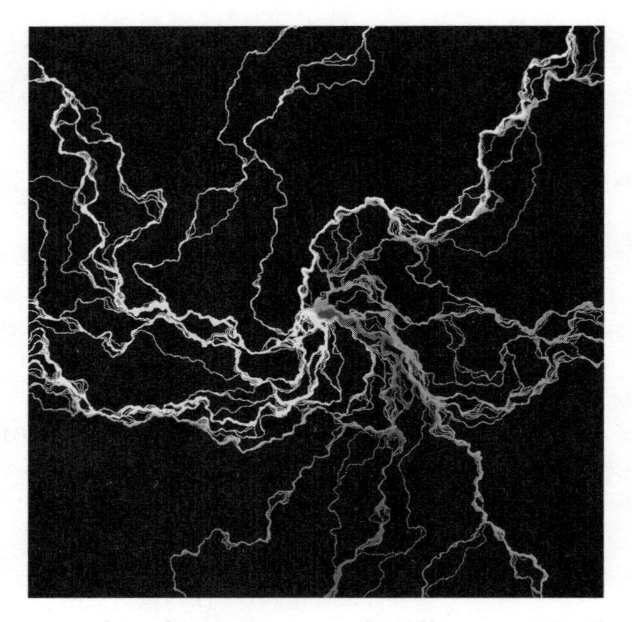

FIGURE 4. Rays with randomness set to a value of kappa = 4/5. Photo by Scott Sheffield. See also color image.

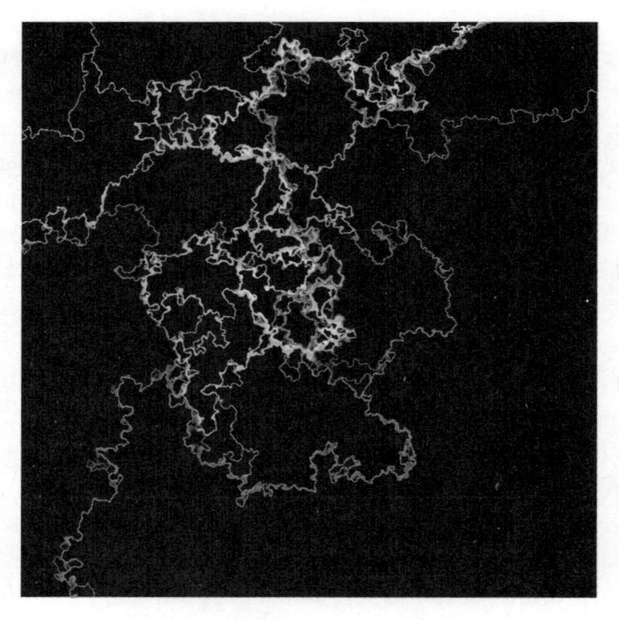

FIGURE 5. Rays with randomness set to a value of kappa = 2. Photo by Scott Sheffield.

Robert Brown observed the random movements of pollen grains suspended in water. The seemingly random motion was caused by individual water molecules bumping into each pollen grain. Later, in the 1920s, Norbert Wiener of MIT gave a precise mathematical description of this process, which is called Brownian motion.

Brownian motion is the "scaling limit" of random walks—if you consider a random walk where each step size is very small, and the amount of time between steps is also very small, these random paths look more and more like Brownian motion. It's the shape that almost all random walks converge to over time.

Two-dimensional random spaces, in contrast, first preoccupied physicists as they tried to understand the structure of the universe.

In string theory, one considers tiny strings that wiggle and evolve in time. Just as the time trajectory of a point can be plotted as a one-dimensional curve, the time trajectory of a string can be understood as a two-dimensional curve. This curve, called a *worldsheet*, encodes the history of the one-dimensional string as it wriggles through time.

"To make sense of quantum physics for strings," said Sheffield, "you want to have something like Brownian motion for surfaces."

For years, physicists have had something like that, at least in part. In the 1980s, physicist Alexander Polyakov, who's now at Princeton University, came up with a way of describing these surfaces that came to be called Liouville quantum gravity (LQG). It provided an incomplete but still useful view of random two-dimensional surfaces. In particular, it gave physicists a way of defining a surface's angles so that they could calculate the surface area.

In parallel, another model, called the Brownian map, provided a different way to study random two-dimensional surfaces. Where LQG facilitates calculations about area, the Brownian map has a structure that allows researchers to calculate distances between points. Together, the Brownian map and LQG gave physicists and mathematicians two complementary perspectives on what they hoped were fundamentally the same object. But they couldn't prove that LQG and the Brownian map were in fact compatible with each other.

"It was this weird situation where there were two models for what you'd call the most canonical random surface, two competing random surface models, that came with different information associated with them," said Sheffield.

Beginning in 2013, Sheffield and Miller set out to prove that these two models described fundamentally the same thing.

The Problem with Random Growth

Sheffield and Miller began collaborating thanks to a kind of dare. As a graduate student at Stanford in the early 2000s, Sheffield worked under Amir Dembo, a probability theorist. In his dissertation, Sheffield formulated a problem having to do with finding order in a complicated set of surfaces. He posed the question as a thought exercise as much as anything else.

"I thought this would be a problem that would be very hard and take 200 pages to solve and probably nobody would ever do it," Sheffield said.

But along came Miller. In 2006, a few years after Sheffield had graduated, Miller enrolled at Stanford and also started studying under Dembo, who assigned him to work on Sheffield's problem as a way of getting to know random processes. "Jason managed to solve this, I was impressed, we started working on some things together, and eventually we had a chance to hire him at MIT as a postdoc," Sheffield said.

In order to show that LQG and the Brownian map were equivalent models of a random two-dimensional surface, Sheffield and Miller adopted an approach that was simple enough conceptually. They decided to see if they could invent a way to measure distance on LQG surfaces and then show that this new distance measurement was the same as the distance measurement that came packaged with the Brownian map.

To do this, Sheffield and Miller thought about devising a mathematical ruler that could be used to measure distance on LQG surfaces. Yet they immediately realized that ordinary rulers would not fit nicely into these random surfaces—the space is so wild that one cannot move a straight object around without the object getting torn apart.

The duo forgot about rulers. Instead, they tried to reinterpret the distance question as a question about growth. To see how this works, imagine a bacterial colony growing on some surface. At first it occupies a single point, but as time goes on it expands in all directions. If you wanted to measure the distance between two points, one (seemingly roundabout) way of doing that would be to start a bacterial colony at one point and measure how much time it took the colony to encompass the other point. Sheffield said that the trick is to somehow "describe this process of gradually growing a ball."

It's easy to describe how a ball grows in the ordinary plane, where all points are known and fixed and growth is deterministic. Random growth is far harder to describe and has long vexed mathematicians. Yet as Sheffield and Miller were soon to learn, "[random growth] becomes easier to understand on a random surface than on a smooth surface," said Sheffield. The randomness in the growth model speaks, in a sense, the same language as the randomness on the surface on which the growth model proceeds. "You add a crazy growth model on a crazy surface, but somehow in some ways it actually makes your life better," he said.

The following images show a specific random growth model, the Eden model, which describes the random growth of bacterial colonies. The colonies grow through the addition of randomly placed clusters along their boundaries. At any given point in time, it's impossible to know for sure where on the boundary the next cluster will appear. In these images, Miller and Sheffield show how Eden growth proceeds over a random two-dimensional surface.

Figure 6 shows Eden growth on a fairly flat—that is, not especially random—LQG surface. The growth proceeds in an orderly way, forming nearly concentric circles that have been color-coded to indicate the time at which growth occurs at different points on the surface.

In subsequent images, Sheffield and Miller illustrate growth on surfaces of increasingly greater randomness. The amount of randomness in the function that produces the surfaces is controlled by a constant, gamma. As gamma increases, the surface becomes rougher—with higher peaks and lower valleys—and random growth on that surface similarly takes on a less orderly form. In Figure 6, gamma is 0.25. In Figure 7, gamma is set to 1.25, introducing five times as much randomness into the construction of the surface. Eden growth across this uncertain surface is similarly distorted.

When gamma is set to the square root of eight-thirds (approximately 1.63), LQG surfaces fluctuate even more dramatically (Figure 8). They also take on a roughness that matches the roughness of the Brownian map, which allows for more direct comparisons between these two models of a random geometric surface.

Random growth on such a rough surface proceeds in a very irregular way. Describing it mathematically is like trying to anticipate minute

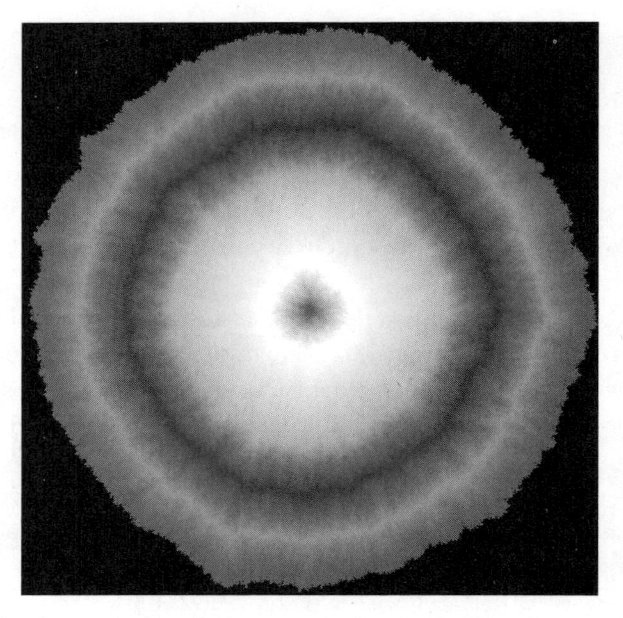

FIGURE 6. Eden growth with gamma equal to 0.25. Photo by Jason Miller.

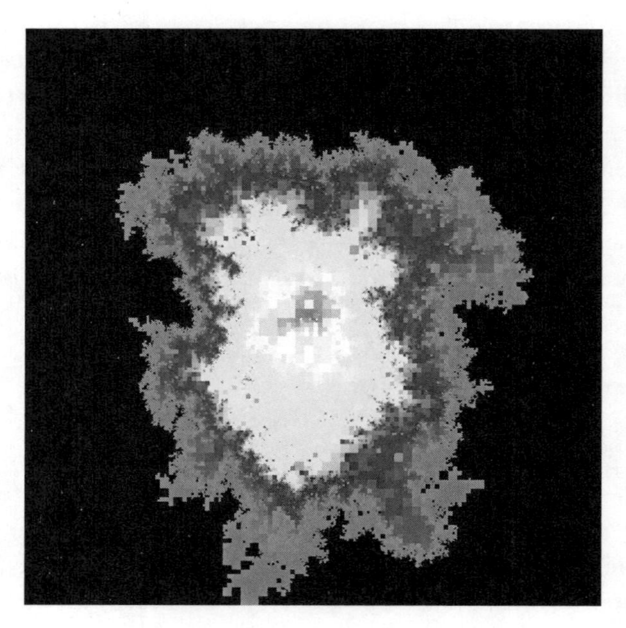

FIGURE 7. Eden growth with gamma equal to 1.25. Photo by Jason Miller.

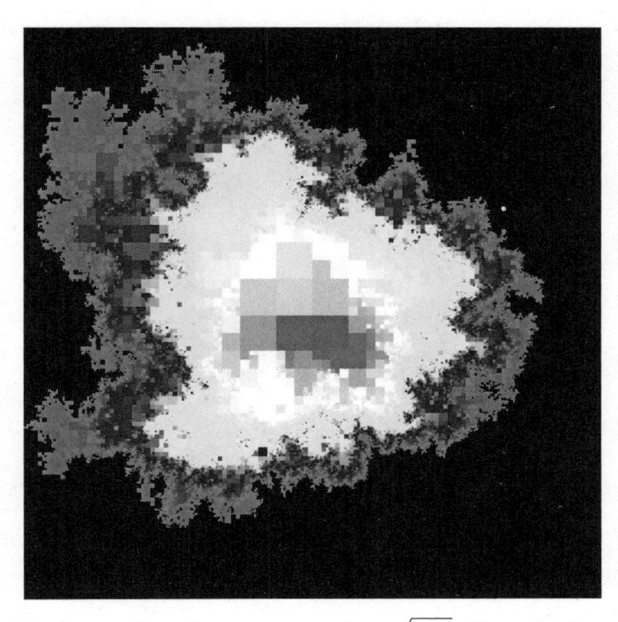

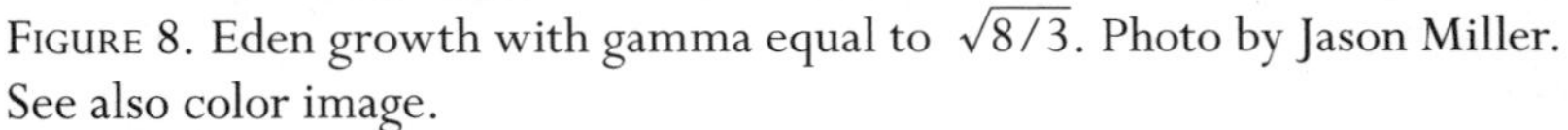

FIGURE 8. Eden growth with gamma equal to $\sqrt{8/3}$. Photo by Jason Miller. See also color image.

pressure fluctuations in a hurricane. Yet Sheffield and Miller realized that they needed to figure out how to model Eden growth on very random LQG surfaces in order to establish a distance structure equivalent to the one on the (very random) Brownian map.

"Figuring out how to mathematically make [random growth] rigorous is a huge stumbling block," said Sheffield, noting that Martin Hairer of the University of Warwick won the Fields Medal in 2014 for work that overcame just these kinds of obstacles. "You always need some kind of amazing clever trick to do it."

Random Exploration

Sheffield and Miller's clever trick is based on a special type of random one-dimensional curve that is similar to the random walk except that it never crosses itself. Physicists had encountered these kinds of curves for a long time in situations where, for instance, they were studying the boundary between clusters of particles with positive and negative spin (the boundary line between the clusters of particles is a one-dimensional

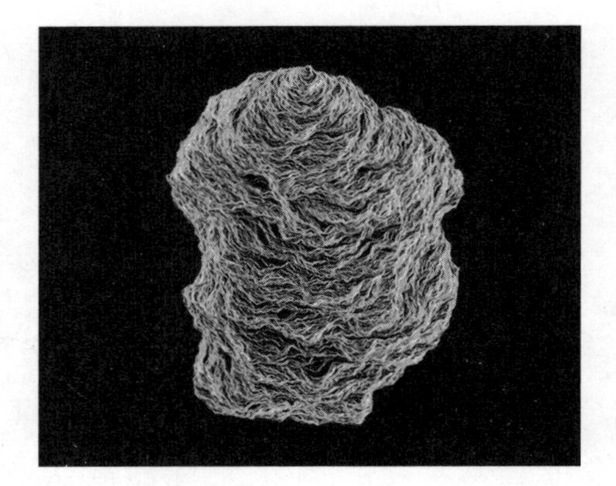

FIGURE 9. An example of an SLE curve. Photo by Jason Miller.

path that never crosses itself and takes shape randomly). They knew these kinds of random, noncrossing paths occurred in nature, just as Robert Brown had observed that random crossing paths occurred in nature, but they didn't know how to think about them in any kind of precise way. In 1999, Oded Schramm, who at the time was at Microsoft Research in Redmond, Washington, introduced the SLE curve (for Schramm–Loewner evolution) as the canonical noncrossing random curve (Figure 9).

Schramm's work on SLE curves was a landmark in the study of random objects. It's widely acknowledged that Schramm, who died in a hiking accident in 2008, would have won the Fields Medal had he been a few weeks younger at the time he'd published his results. (The Fields Medal can be given only to mathematicians who are not yet 40.) As it was, two people who worked with him built on his work and went on to win the prize: Wendelin Werner in 2006 and Stanislav Smirnov in 2010. More fundamentally, the discovery of SLE curves made it possible to prove many other things about random objects.

"As a result of Schramm's work, there were a lot of things in physics they'd known to be true in their physics way that suddenly entered the realm of things we could prove mathematically," said Sheffield, who was a friend and collaborator of Schramm's.

For Miller and Sheffield, SLE curves turned out to be valuable in an unexpected way. In order to measure distance on LQG surfaces, and thus show that LQG surfaces and the Brownian map were the same,

FIGURE 10. An SLE curve with kappa equal to 0.5. Photo by Jason Miller. See also color image.

they needed to find some way to model random growth on a random surface. SLE proved to be the way.

"The 'aha' moment was [when we realized] you can construct [random growth] using SLEs and that there is a connection between SLEs and LQG," said Miller.

SLE curves come with a constant, kappa, which plays a similar role to the one gamma plays for LQG surfaces. Where gamma describes the roughness of an LQG surface, kappa describes the "windiness" of SLE curves. When kappa is low, the curves look like straight lines. As kappa increases, more randomness is introduced into the function that constructs the curves and the curves turn more unruly, while obeying the rule that they can bounce off of, but never cross, themselves. Figure 10 is an SLE curve with kappa equal to 0.5, and Figure 11 is an SLE curve with kappa equal to 3.

Sheffield and Miller noticed that when they dialed the value of kappa to 6 and gamma up to the square root of eight-thirds, an SLE curve drawn on the random surface followed a kind of exploration process. Thanks to works by Schramm and by Smirnov, Sheffield and Miller knew that when kappa equals 6, SLE curves follow the trajectory of a kind of "blind explorer" who marks her path by constructing a trail as she goes. She moves as randomly as possible, except that whenever she bumps into a piece of the path she has already followed, she turns away from that piece to avoid crossing her own path or getting stuck in a dead end.

FIGURE 11. An SLE curve with kappa equal to 3. Photo by Jason Miller. See also color image.

"[The explorer] finds that each time her path hits itself, it cuts off a little piece of land that is completely surrounded by the path and can never be visited again," said Sheffield.

Sheffield and Miller then considered a bacterial growth model, the Eden model, that had a similar effect as it advanced across a random surface: It grew in a way that "pinched off" a plot of terrain that, afterward, it never visited again. The plots of terrain cut off by the growing bacteria colony looked exactly the same as the plots of terrain cut off by the blind explorer. Moreover, the information possessed by a blind explorer at any time about the outer unexplored region of the random surface was exactly the same as the information possessed by a bacterial colony. The only difference between the two was that while the bacterial colony grew from all points on its outer boundary at once, the blind explorer's SLE path could grow only from the tip.

In a paper posted online in 2013, Sheffield and Miller imagined what would happen if, every few minutes, the blind explorer were magically transported to a random new location on the boundary of the territory she had already visited. By moving all around the boundary, she would be effectively growing her path from all boundary points at once, much like the bacterial colony. Thus they were able to take something they could understand—how an SLE curve proceeds on a random surface—and show that with some special configuring, the curve's evolution exactly described a process they hadn't been able to understand, random growth. "There's something special about the relationship between SLE and growth," said Sheffield. "That was kind of the miracle that made everything possible."

The distance structure imposed on LQG surfaces through the precise understanding of how random growth behaves on those surfaces exactly matched the distance structure on the Brownian map. As a result, Sheffield and Miller merged two distinct models of random two-dimensional shapes into one coherent, mathematically understood fundamental object.

Turning Randomness into a Tool

Sheffield and Miller have already posted the first two papers in their proof of the equivalence between LQG and the Brownian map on the scientific preprint site arxiv.org; they intend to post the third and final paper later this summer. The work turned on the ability to reason across different random shapes and processes—to see how random noncrossing curves, random growth, and random two-dimensional surfaces relate to one another. It's an example of the increasingly sophisticated results that are possible in the study of random geometry.

"It's like you're in a mountain with three different caves. One has iron, one has gold, one has copper—suddenly you find a way to link all three of these caves together," said Sheffield. "Now you have all these different elements you can build things with and can combine them to produce all sorts of things you couldn't build before."

Many open questions remain, including determining whether the relationship between SLE curves, random growth models, and distance measurements holds up in less-rough versions of LQG surfaces than the one used in the current paper. In practical terms, the results by Sheffield and Miller can be used to describe the random growth of real phenomena like snowflakes, mineral deposits, and dendrites in caves, but only when that growth takes place in the imagined world of random surfaces. It remains to be seen whether their methods can be applied to ordinary Euclidean space, like the space in which we live.

An "Infinitely Rich" Mathematician Turns 100

Siobhan Roberts

At the Hotel Parco dei Principi in Rome, in September of 1973, the Hungarian mathematician Paul Erdős approached his friend Richard Guy with a request. He said, "Guy, veel you have a coffee?" It cost a dollar, a small fortune for a professor of mathematics at the hinterland University of Calgary who was not much of a coffee drinker. Yet, as Guy later recalled—during a memorial talk following Erdős's death at age 83 two decades ago—he was curious why the great man had sought him out.

Guy and Erdős were in the Eternal City for an international colloquium on combinatorial theory, so Erdős—who sustained himself with espresso and other stimulants, worked on math problems 19 hours a day, and in his lifetime published in excess of 1,500 papers with more than 500 collaborators—most likely had another problem on the go. When they sat with their coffee, he said, "Guy, you are eenfeeneeteley reech. Lend me 100 dollars."

"I was amazed," recounted Guy. "Not so much at the request but rather at my ability to satisfy it. Once again, Erdős knew me better than I know myself. Ever since then, I've realized that I'm infinitely rich: Not just in the material sense that I have everything I need, but infinitely rich in spirit in having mathematics and having known Erdős."

Today, at the newly minted age of 100, Richard Guy still considers himself "eenfeeneeteley reech." A lifelong mountain climber and environmentalist, he spent his centenary day hiking at—not up—Mount Assiniboine, the so-called Matterhorn of the Rockies, arriving at the trailhead via helicopter. This escapade followed a summer of mathematically motivated travel, including Columbus, Ohio, for the Mathematical Association of America's annual "MathFest," where during an early birthday celebration Guy led a sing-along to the tune of "My Bonnie Lies Over the Ocean;"

> Bring back! Oh, bring back!
> Bring back that Gee-om-met-tree to me!
> Bring back! Bring back!
> Bring back that Gee-om-met-tree!

"He is a legend, an icon, a classic, but not a fossil," said Marjorie Senechal, editor-in-chief of *The Mathematical Intelligencer,* who still asks Guy to referee papers and often consults his good judgment on any number of matters. "He knows everything," she said.

As a boldface name within the mathematical community, Guy is right up there with Oxford's Andrew Wiles, who solved Fermat's Last Theorem, and the Fields Medalist Terence Tao, the "Mozart of maths" at the University of California, Los Angeles. He achieved this status by working away as a self-described amateur, though he pushes the boundaries of that definition. Despite his lack of a Ph.D. (and of any pretense), at the age of 50 he became a professor at the University of Calgary, but before that he was a high school teacher. The Princeton mathematician Manjul Bhargava, a 2014 Fields Medalist, grew up listening to an uncle talk about his fantastic calculus teacher in Delhi, one Richard Guy. Later, Bhargava discovered the wonderful works of a famed number theorist, also named Richard Guy. "Only many years after that did I realize these were the same Richard Guys!" he said.

Guy is known for finding the crucial "glider" in John Horton Conway's Game of Life, a universal cellular automaton that produces rich complexity from a very simple rule set—the glider was a wiggling, skittering animal, of sorts, gliding its way diagonally across the Life board. And Guy contributed to the canon of mathematical folklore with a tongue-in-cheek paper, published in 1988, titled, "The Strong Law of Small Numbers," wherein he cautions: "Superficial similarities spawn spurious statements. Capricious coincidences cause careless conjectures . . ."

He is perhaps most famous, however, for posing problems—not so much solving problems (he does that, too); rather collecting and curating problems, with signature perspicuity, and propagating them in books such as, *Unsolved Problems in Number Theory.* "Erdős was the main inspiration," Guy told me. First published in 1981, the third edition is still in print. "You can tell a mathematician by the questions they ask, not by the answers they give," said Andrew Granville, of the University College London. "Some of Erdős's questions that I'm sure Richard collected over the years I thought were quite daft when he asked them, but they turned out to be exactly the right questions."

"That's how new mathematics is discovered, by looking at statements of problems that we can understand, but that we cannot say why they are or are not true," said Carl Pomerance, a number theorist at Dartmouth College. He and Guy are working on an ancient problem that intrigued Pythagoras and involves so-called "amicable numbers." The numbers 220 and 284 are amicable numbers, because the proper divisors of 220 add up to 284, and the proper divisors of 284 add up to 220. Finding all the divisors of a number is a hard problem, with tangential applications. "The problem of factoring big numbers is the cornerstone to security on the Internet, public-key crypto systems, and that problem cut its teeth on this old problem of Pythagoras," said Pomerance.

Beguiling problems—even the trivial and goofy ones that may or may not transform into serious and profound problems like the Riemann Hypothesis or Fermat's Last Theorem—are the lifeblood of mathematics. As Guy noted in the preface to his first edition of *Unsolved Problems,* "Mathematics is kept alive . . . by the appearance of a succession of unsolved problems both from within mathematics itself and from the increasing number of disciplines where it is applied." He made note of the problem first posed by the French lawyer and amateur mathematician Pierre de Fermat in 1637. But in 1994, the year of the second edition, Wiles announced his proof, thus transforming Fermat's conjecture into a theorem. "This book is perpetually out of date," Guy noted, with pleasure, in the new preface. "That is one of the many beauties of mathematics," he told me—"that it's perpetually out of date!"

Wiles went into mathematics having been romanced since boyhood by Fermat's unsolved problem. But in his view, a young mathematician must be discriminating in tackling problems, taking care to avoid trivial distractions that lead nowhere. "I think it's very important to have problems that are really connected to the development of a field," he said. Others take a more inclusive and optimistic view. "I liken problems to little acorns," said Ron Graham, of the University of California, San Diego. "You plant them, and maybe nothing happens, or maybe eventually a giant oak tree sprouts. You never know." Guy's approach treats problems like games—he plays with the problem, preferring to work from first principles and by dint of cleverness, rather than with a black box of doodads previously engineered by somebody else. "[Guy's] like an artist, with hands on the canvas," said Wiles.

Guy in fact got his start gaming, as a chess problemist. While a student at Cambridge in the 1930s, he spent too much time composing endgame problems and achieved only a second-class degree. Then, as Guy likes to say, he enjoyed a checkered career as a meteorologist in Iceland and Bermuda, and as a teacher in Singapore and India. In Singapore in 1960, Erdős paid him a visit—Erdős being the problem poser *par excellence*, according to Guy. The itinerant Hungarian traveled from university to university, delivering some 40 talks a year, posing his problems, occasionally offering cash prizes for solutions, sometimes 25 dollars, other times thousands. "He stayed with me and gave me two or three of his problems," Guy once recalled. "I made some progress in each of them. This gave me encouragement, and I began to think of myself as possibly being something of a research mathematician, which I hadn't done before."

Technically, Guy retired in 1982, though he forgot to stop going into his office, where he sits, door open, nearly every day. He sends out a mass e-mail notification when he's due to be away. Last summer, during a whirlwind trip to New York City, he delivered a talk at MoMATH on the little known fact that "A Triangle Has Eight Vertices," and he took a meeting (thanks to the efforts of maestro magician and matchmaker Mark Mitton) with the Russian chess grandmaster Garry Kasparov. The pair were joined by Conway and Elwyn Berlekamp, Guy's coauthors on the bestselling classic, "Winning Ways for Your Mathematical Plays."

And just before his visit to MathFest, he attended the annual meeting of the Canadian Number Theory Association. Toward the end of a session feting the man and his work, Guy took the floor and fielded questions. "Give me a problem," he said, "and I'll *unsolve* it.

Inverse Yogiisms

Lloyd N. Trefethen

Berra Backwards

The great New York Yankees catcher Yogi Berra died in September 2015. Berra was famous for his quirky sayings, like these:

"It ain't over till it's over."
"When you come to a fork in the road, take it."
"It gets late early out there."
"A nickel ain't worth a dime anymore."
"I always thought that record would stand until it was broken."
"You wouldn't have won if we'd beaten you."
"Nobody goes there anymore, it's too crowded."

Maybe Berra never said half the things he said, but that's not the point. We have here a brand of malapropisms that people have been enjoying for years.

It's pretty easy to spot the trick that animates these quips. Yogiisms are statements that, if taken literally, are meaningless or contradictory or nonsensical or tautological—yet nevertheless convey something true. It's a clever twist that gets us smiling and paying attention. If you like, you could argue that literature and art sometimes use the same device. A Yogiism is like a Picasso painting, you could say, messing with reality in a manner that catches our interest and still conveys a truth.

But I want to stay with words and their meanings. I think Yogiisms hold a special lesson for mathematicians, because our characteristic pitfall, I propose, is the *inverse Yogiism*: the statement that is literally true, yet conveys something false.

At some level, we're all well aware that saying useless true things is an occupational hazard. Just think of that joke about the people lost in a hot air balloon who shout, "Where are we?" to a man on the ground.

"You're in a balloon!" the mathematician answers. (I have heard this joke far too often.)

So we all know in a general way about our habit of taking things literally. My proposal is that this phenomenon is more important than we may realize, and the notion of an inverse Yogiism can help us focus on it. Inverse Yogiisms in mathematics, and in science more generally, can impede progress sometimes for generations. I will describe two examples from my own career and then mention a third topic, more open-ended, that may be a very big example indeed.

Faber's Theorem on Polynomial Interpolation

The early 1900s was an exciting time for the constructive theory of real functions. The old idea of a function as given by a formula had broadened to include arbitrary mappings defined pointwise, and connecting the two notions was a matter of wide interest. In particular, mathematicians were concerned with the problem of approximating a continuous function f defined on an interval such as $[-1,1]$ by a polynomial p. Weierstrass's theorem of 1885 had shown that arbitrarily close approximations always exist, and by 1912, alternative proofs had been published by Picard, Lerch, Volterra, Lebesgue, Mittag-Leffler, Fejér, Landau, de la Vallée Poussin, Jackson, Sierpiński, and Bernstein.

How could polynomial approximations be constructed? The simplest method would be interpolation by a degree-n polynomial in a set of $n + 1$ distinct points in $[-1,1]$. Runge showed in 1900 that interpolants in equally spaced points will not generally converge to f as $n \to \infty$, even for analytic f. On the other hand, Chebyshev grids with their points clustered near ± 1 do much better. Yet around 1912, it became clear to Bernstein, Jackson, and Faber that no system of interpolation points could work for all functions. The famous result was published by Faber in 1914, and here it is in his words and notation (translated from [4]).

> **Faber's Theorem.** *There does not exist any set E of interpolation points $x_i^{(n)}$ in $s = (-1,\ 1)$ $(n = 1,2 \ldots ;\ i = 1,2 \ldots n + 1)$ with the property that every continuous function F(x) in s can be represented as the uniform limit of the degree-n polynomials taking the same values as F for $x = x_i^{(n)}$.*

The proof nowadays (though not yet in 1914) makes elegant use of the uniform boundedness principle.

Faber's theorem is true, and moreover, it is beautiful. Let me now explain how its influence has been unfortunate.

The field of numerical analysis took off as soon as computers were invented, and the approximation of functions was important in every area. You might think that polynomial interpolation would have been one of the standard tools from the start, and to some extent this was the case. However, practitioners must have often run into trouble when they worked with polynomial interpolants—usually because of using equispaced points or unstable algorithms, I suspect—and Faber's theorem must have looked like some kind of explanation of what was going wrong. The hundreds of textbooks that soon began to be published fell into the habit of teaching students that interpolation is a dangerous technique, not to be trusted. Here are some illustrations.

Isaacson and Keller, *Analysis of Numerical Methods* (1966), p. 275:

> It is not generally true that higher degree interpolation polynomials yield more accurate approximations.

Kahaner, Moler, and Nash, *Numerical Methods and Software* (1989), p. 94:

> Polynomial interpolants rarely converge to a general continuous function.

Kincaid and Cheney, *Numerical Analysis* (1991), p. 319:

> The surprising state of affairs is that for most continuous functions, the quantity $\|f - p_n\|_\infty$ will not converge to 0.

Stoer and Bulirsch, *Introduction to Numerical Analysis* (1993), p. 51:

> It should not be assumed that finer and finer samplings of the function f will lead to better and better approximations through interpolation.

Stewart, *Afternotes on Numerical Analysis* (1996), p. 153:

> Unfortunately, there are functions for which interpolation at the Chebyshev points fails to converge.

Gautschi, *Numerical Analysis: An Introduction* (1997), p. 79:

> It is not possible, therefore, to conclude . . . that Lagrange interpolation converges uniformly on $[a,b]$ for any continuous function, not even for judiciously selected nodes; indeed, one knows that it does not.

Quarteroni, Sacco, and Saleri, *Numerical Mathematics* (2000), p. 331:
> Thus, polynomial interpolation does not allow for approximating *any* continuous function. . . .

What a load of inverse Yogiisms! Statements like these, which appear in so many of the textbooks, give entirely the wrong impression. In fact, polynomial interpolation in Chebyshev points is a powerful and reliable method for approximation of functions. The Chebfun software system routinely works with degrees in the thousands [2].

The flaw in the logic is that Faber's theorem says nothing if f is smooth [7]. If f is Lipschitz continuous, that is more than enough to guarantee convergence of interpolants in Chebyshev points, and if it has a kth derivative of bounded variation, the error in the degree-n interpolant is of size $O(n^{-k})$. If f is analytic, the convergence is at a geometric rate $O(\rho^{-n})$, $\rho > 1$. Moreover, there are methods for computing these interpolants that are fast and numerically stable, notably the so-called barycentric interpolation formula.

So the idea that polynomial interpolants can't be trusted is a myth: a myth that has drawn strength from an impeccable theorem. Make sure your functions are Lipschitz continuous or better, as is easily done in almost any application, and Faber's theorem ceases to be applicable. In fact, polynomial interpolation in Chebyshev points has the same power and robustness as discrete Fourier analysis, to which it is essentially equivalent. We must hope that the numerical analysis textbooks of future generations will begin to tell students this.

Squire's Theorem on Hydrodynamic Instability

We now move from numerical analysis to one of the oldest problems of fluid mechanics. Consider the idealized *plane Poiseuille flow* of a Newtonian liquid or gas in an infinite channel between two flat plates. (The mathematics is similar for other geometries such as a circular pipe, as investigated by Reynolds in 1883.) The flow is governed by the Navier–Stokes equations, and the key parameter is the Reynolds number, Re, a nondimensionalized velocity.

Will the flow be laminar or turbulent? At low values of Re, it is the laminar solution one sees in the laboratory, a smooth parallel downstream flow with a fixed velocity profile in the shape of a parabola. At high Re, though the laminar flow remains a mathematically valid solution of the equations, what one sees in the lab are the chaotic whirls and eddies of turbulence. Now if this is so, it seems clear that for high Re, the laminar flow must be unstable in the sense that small perturbations of that flow may get amplified. In an analysis going back to Orr and Sommerfeld in 1907–1908, one makes this precise by linearizing the equations about the laminar solution, obtaining a linear operator L_{Re} that governs the evolution of infinitesimal perturbations. If L_{Re} has an eigenfunction corresponding to an eigenvalue in the right half of the complex plane, this represents an infinitesimal perturbation that can grow exponentially, so the flow should be unstable; and if not, it should be stable.

This brings us to the elegant result published by Herbert Brian Squire in 1933. The geometry of our planar domain is 3D, with variables x (streamwise), y (perpendicular to the plates), and z (spanwise). Analyzing the linearized operator for this 3D flow is going to be complicated. Squire's theorem, however, tells us we can ignore the z direction and just do a 2D analysis. Here is the statement from his original paper [6].

> **Squire's Theorem.** *Any instability which may be present for three-dimensional disturbances is also present for two-dimensional disturbances at a lower value of Reynolds' number.*

The influence of Squire's theorem can be seen all across the literature of mathematical fluid mechanics. Whenever you see an analysis involving the famous Orr–Sommerfeld equation, the authors have probably taken a 3D flow problem and reduced it to 2D. For the plane Poiseuille configuration, the theory tells us that the 2D instability sets in at a critical Reynolds number $\mathrm{Re}_c \approx 5{,}772.22$, a threshold first calculated accurately by Orszag. For $\mathrm{Re} < \mathrm{Re}_c$, we should expect stability and laminar flow, and for $\mathrm{Re} > \mathrm{Re}_c$, instability and turbulence.

Here are summaries from some books.

Lin, *The Theory of Hydrodynamic Stability* (1967), p. 27:

> We shall now show, following Squire (1933), that the problem of three-dimensional disturbances is actually equivalent to a two-dimensional problem at a *lower* Reynolds number.

Tritton, *Physical Fluid Dynamics* (1977), p. 220:

> . . . there is a result, known as Squire's theorem, that in linear stability theory the critical Reynolds number for a

two-dimensional parallel flow is lowest for two dimensional perturbations. We may thus restrict attention to these.

Drazin and Reid, *Hydrodynamic Stability* (1981), p. 155:

Squire's theorem. To obtain the minimum critical Reynolds number it is sufficient to consider only two-dimensional disturbances.

Friedlander and Serre, eds., *Handbook of Mathematical Fluid Dynamics,* v. 3 (2002), p. 248:

THEOREM 1.1 (Squire's theorem, 1933). To each unstable three-dimensional disturbance there corresponds a more unstable two-dimensional one.

Sengupta, *Instabilities of Flows and Transition to Turbulence* (2012), p. 82:

In a two-dimensional boundary layer with real wave numbers, instability appears first for two-dimensional disturbances.

These and other sources are in agreement on a very clear picture, and only one thing is amiss: the picture is wrong! In the laboratory, observed structures related to transition to turbulence are almost invariably three-dimensional. Moreover, it is difficult to spot any change of flow behavior at $\text{Re} \approx 5{,}772.22$. For $\text{Re} < \text{Re}_c$, many flows are turbulent when we expect them to be laminar. For $\text{Re} > \text{Re}_c$, many flows are laminar when we expect them to be turbulent. What is going on?

The flaw in the logic is that eigenmodal analysis applies in the limit $t \to \infty$, whereas the values of t achievable in the laboratory rarely exceed 100. (Thanks to the nondimensionalization, t is related to the *length* of a flow apparatus relative to its width.) Consequently, high-Reynolds number flows normally do not become turbulent in an eigenmodal fashion [8]. On the one hand, the exponential growth rates of unstable eigenfunctions, known as Tollmien–Schlichting waves, are typically so low that in a laboratory setup they struggle to amplify a perturbation by even a factor of 2. This is why laminar flow is often observed with $\text{Re} \gg \text{Re}_c$. On the other hand, much faster transient amplification mechanisms are present for 3D perturbations, even for $\text{Re} \ll \text{Re}_c$. The perturbations involved are not eigenfunctions, and in principle they would die out as $t \to \infty$ if they started out truly infinitesimal: Squire's theorem is, of course, literally true. But the transient growth of 3D perturbations is so substantial that in a real flow, small finite disturbances may quickly be raised to a level where nonlinearities kick in. In assuring us that the most dangerous disturbances are two-dimensional,

Squire's theorem has told us exactly the wrong place to look for hydrodynamic instability.

P = NP?

The most famous problem in computer science, which is also one of the million-dollar Clay Millennium Prize Problems, is the celebrated question "P = NP?" This puzzle remains unresolved half a century after it was first posed by Cook and Levin in 1971.

Some computational problems can be solved by fast algorithms, and others only by slow ones. One might expect a continuum of difficulty, but the unlocking observation was that there is a gulf between *polynomial time* and *exponential time* algorithms. Inverting an $n \times n$ matrix, say, can be done in $O(n^3)$ operations or less, so we deal routinely with matrices with dimensions in the thousands. Finding the shortest route for a salesman visiting n cities, on the other hand, requires C^n operations for some $C > 1$ by all algorithms yet discovered. As $n \to \infty$, the difference between n^c and C^n looks like a clean binary distinction. And there is a great class of thousands of problems, the NP-complete problems, that have been proved to be equivalent in the sense that all of them can be solved in polynomial time or none of them can—and nobody knows which.

It's an extraordinary gap in our knowledge. If I may pick two mathematical mysteries that I hope will be resolved before I die, they are the Riemann hypothesis and "P=NP?" It is so crisp, and so important!

Yet Yogi Berra seems to be looking over our shoulders. Computers are millions of times more powerful than they were in 1971, increasing the tractable size of n in every problem known to man, so one might expect that the gulf between P and the best known algorithms for NP, which seemed significant already in 1971, should have opened up by now to a canyon so deep we can't see its bottom. Yet in the event, nothing so straightforward has happened. Some NP-complete problems still defeat us. Others are easily solvable in many instances. For example, one of the classic NP-complete problems is "SAT," involving the satisfiability of Boolean expressions. SAT solvers have become so powerful that they are now a standard computational tool, solving problem instances of scales in the thousands and even millions [5]. Surprisingly powerful methods have been developed for other NP-complete problems too, including integer programming [1] and the traveling salesman problem itself [3].

So is there a logical flaw in "P=NP?" as with Faber's theorem and Squire's theorem? I would not go so far as to say this, but it is certainly the case that, once again, the precision of a mathematical formulation has encouraged us to think the truth is simpler than it is. A typical NP-complete problem measures complexity by the *worst case time* required to deliver the *optimal solution*. Experience has shown that in practice, both ends of this formulation are negotiable. For some NP-complete problems, like SAT, the worst case indeed looks exponential but in practice it is rare for a problem instance to come close to the worst case. For others, like the "max-cut" problem, it can be proved that even in the worst case one can solve the problem in polynomial time, if one is willing to miss the optimum by a few percent (for max-cut, 13 percent is enough). A field of approximation algorithms has grown up that develops algorithms of this flavor [9]. Often these algorithms rely on tools of continuous mathematics to approximate problems formulated discretely. Indeed, the whole basis of "P=NP?" is a discrete view of the world, and the distracting sparkle of this great unsolved problem may have delayed the recognition that often, continuous algorithms have advantages even for discrete problems. As scientists we must always simplify the world to make sense of it; the challenge is to not get trapped by our simplifications.

Coda

When we say something precisely and even prove that it's true, we open ourselves to the risk of inverse Yogiisms. Would it be better if mathematicians didn't try so hard to be precise? Certainly not! Rigorous theorems are the pride of mathematics, which enable this unique subject to advance from one century to the next. The point is only that we must always strive to examine a problem from different angles, to think widely about its context as well as technically about its details. Or as Yogi put it,

"Sometimes you can see a lot just by looking."

Acknowledgments

Many people made good suggestions as I was writing this essay. I would like particularly to acknowledge Tadashi Tokieda and David Williamson.

References

Books cited in the quotations are readily tracked down from the information given there; I have quoted from the earliest editions I had access to. Other references are as below.

[1] R. E. Bixby, A brief history of linear and mixed-integer programming computation, *Documenta Mathematica* (2012), 107–121.

[2] Chebfun software project, www.chebfun.org.

[3] W. J. Cook, The Traveling Salesman Problem, www.math.uwaterloo.ca/tsp.

[4] G. Faber, Über die interpolatorische Darstellung stetiger Funktionen, *Jber. Deutsch. Math. Verein.* 23 (1914), 192–210.

[5] D. E. Knuth, *Satisfiability*, v. 4 fascicle 6 of The Art of Computer Programming, 2015.

[6] H. B. Squire, On the stability for three-dimensional disturbances of viscous fluid flow between parallel walls, *Proc. Roy. Soc. Lond. A* 142 (1933), 621–628.

[7] L. N. Trefethen, *Approximation Theory and Approximation Practice*, SIAM, 2013.

[8] L. N. Trefethen, A. E. Trefethen, S. C. Reddy, and T. A. Driscoll, Hydrodynamic stability without eigenvalues, *Science* 261 (1993), 578–584.

[9] D. P. Williamson and D. B. Shmoys, *The Design of Approximation Algorithms*, Cambridge U. Press, 2011.

Ramanujan in Bronze

Gerald L. Alexanderson, with Contributions from Leonard F. Klosinski

Srinivasa Ramanujan's name is one of the best known in the history of mathematics. The romantic story of someone born into a family of modest means, raised in a small city in southern India, and who goes on to become one of the greatest mathematicians in all of recorded history has been told again and again. So beyond giving a memory-refreshing outline of the story, I shall try to avoid repeating too many well-known facts.

The young Ramanujan learned some mathematics by reading a few available but long out-of-date mid-19th century English books, one by G. S. Carr (*A Synopsis of Elementary Results in Pure and Applied Mathematics*), a collection of formulas and statements of theorems, seldom with any proofs. He then proceeded to collect in notes amazing conjectures while learning, largely on his own, whole parts of number theory and analysis. He did not survive long in college because he was not interested in most of the required curriculum. He only cared about mathematics. In 1911–1912, he and other Indian mathematicians started sending his discoveries to prominent mathematicians in England, but these were essentially ignored until he sent some of his work to the greatest English mathematician of the time, G. H. Hardy, who recognized that Ramanujan had rediscovered some known but significant results, had guessed wrong in some cases, but had also come up with conjectures that were startlingly deep, important, and unknown up to that time [15]. Hardy arranged to have Ramanujan come to England, which he did in 1914, against the wishes of his mother, who on religious (and probably personal) grounds did not want him to leave India. At Cambridge, he worked with Hardy, J. E. Littlewood, G. N. Watson, and others until his health failed for reasons that are not entirely clear. Tuberculosis has been mentioned, along with the foul winter weather in England, which contrasted with the warm climate of southern India. Being a Brahmin, he was a vegetarian and therefore had great difficulty in finding agreeable food that neither contained meat nor animal fats used in the preparation. There are various other more exotic conjectures, but in the end we do not know for sure what all contributed to his early death in 1920, only months after he returned to India.

One of Ramanujan's best-known discoveries was an exact formula for calculating the number of partitions, $p(n)$, of a positive integer n. This problem involves calculating the number of ways of writing n as a sum of positive integers (including n itself) and ignoring the order in which the numbers appear. For example, it is easy to see that; $p(5) = 7$. In the mid-18th century, Euler had found a recursion formula to calculate values of $p(n)$, but until Ramanujan there was no direct formula known. Hardy and Ramanujan, using their famous "circle method," found the exact formula for $p(n)$, "an achievement undertaken and mostly completed by G. H. Hardy and S. Ramanujan and fully completed and perfected by H. Rademacher. . . . This unbelievable identity [which we will not quote here] wherein the left-hand side is the humble arithmetic function $p(n)$ and the right-hand side is an infinite series involving π, square roots, complex roots of unity, and derivatives of hyperbolic functions, provides not only a theoretical formula for $p(n)$ but also a formula which admits relatively rapid computation" [4, pp. 68–70]. It's a stunning formula, providing, for example, the value of $p(200)$: 3,972,999,029,388.

Ramanujan's notes were filled with extraordinary formulas and observations that often went far beyond what contemporaries had observed, let alone proved. People have claimed, certainly correctly, that he was almost without equal as a mathematical genius, and some have gone on to say that in some ways he went beyond Einstein—everybody's idea of what a mathematical genius should be (and should look like). But there was a difference. Einstein was basing his startling observations on a solid background in mathematics and physics at first-class places, Zurich's Polytechnic (ETH) and the University in Berlin. Ramanujan, by contrast, was largely self-educated.

The narrative, beyond a few stories published in India during Ramanujan's early years, essentially begins with G. H. Hardy's obituary essay in the *Proceedings of the London Mathematical Society* and the *Proceedings of the Royal Society* (reprinted in [8, pp. xl–lviii]), where Hardy tells of Ramanujan's arrival in England where his work dazzled the mathematical community. (My own copy of this book is the one Hardy gave to Pólya at the time of publication.) Within a few years, Ramanujan was being honored with membership in the Royal Society of London and election as Fellow of Trinity College, Cambridge, both in 1918. In 1940, Hardy wrote about him again in [9], an essay on his life and his work in England.

In 1991, Robert Kanigel published a splendid biography, *The Man Who Knew Infinity* [11], which probably remains the best source of information on Ramanujan's life. A movie with the same name as the Kanigel book was released in April 2016 in the United States and received widespread and positive reviews. The cast was star-studded—Hardy is played by Oscar-winning actor Jeremy Irons, Ramanujan by Dev Patel, and lesser roles by Stephen Fry and Jeremy Northam. It was largely filmed at Cambridge and in India. An earlier film on Ramanujan, *The Man Who Loved Numbers*, was shown on public television in the United States in 1988 as part of the *NOVA* series. It was a beautiful film and contained a memorable interview with Ramanujan's wife, Janaki Ammal, who was still alive when the film was made. She had married Ramanujan at the age of nine in a marriage arranged by his mother. The couple lived together for a rather short time before he left for England, so they had little time together until he came back to India and worked on a manuscript known later as the "lost notebooks" in the months remaining before his death. In the film, Janaki is interviewed, and this tiny, frail woman, then in her 90s, said, "All I can tell you is that day and night he worked on sums. He didn't do anything else. He wasn't interested in anything else. He wouldn't stop work even to eat. We had to make rice balls for him and placed them in the palm of his hand. Isn't that extraordinary?" Through history, there must have been many spouses of mathematicians who would sympathize with this woman as they wondered about how their husbands or wives spent their time, "doing their sums."

As recently as 2007 David Leavitt, a writer who had earlier ventured into mathematical biography with his *The Man Who Knew Too Much: Alan Turing and the Invention of the Computer* [14], published a novel about Ramanujan, *The Indian Clerk: A Novel* [13]. Though it was awarded a favorable front-page review in the *New York Times Sunday Book Review* section, it was not to everyone's taste. The great librarian and book expert, Lawrence

Clark Powell, once remarked: "I believe a good work of fiction about a place is a better guide than a bad work of fact" [12, p. 16]. That might in some cases be true, but if one substitutes "person" for "place," it's risky. Leavitt had gotten into trouble with an earlier work, *While England Sleeps*, a novel that the English poet Stephen Spender claimed was based on his memoirs, and he charged that Leavitt misconstrued the facts. So Spender sued. Leavitt made the mistake of publishing his book while Spender was still alive. The case was settled out of court, but the book had to be revised and reissued by the publisher. In the case of *The Indian Clerk*, the cast members were no longer alive and able to object. It raises a serious question about the wisdom of writing historical fiction. It can so easily change into fictional history and spread misinformation. I did not finish reading Leavitt's book because I feared that in time I would come to assume that conversations reported in the book actually took place, when in fact they almost certainly did not.

On the other hand, there was quite a splendid play about Ramanujan, *Partition*, written by Ira Hauptman and produced at the Aurora Theatre in Berkeley in 2003. The principal characters were Ramanujan, Hardy, a fictional classicist named Billington, the Hindu goddess Namagiri of Namakkal (in the play clad in a colorful and elegant sari), and Pierre de Fermat. Now that's clearly fiction because Fermat lived in a different century, but it made sense to include Fermat in a piece of theater because the story worked in a fictional interest on the part of Ramanujan in solving Fermat's Last Theorem [13]. Ramanujan told people that he received some of his ideas from the family deity, the goddess Namagiri, who came to him in his sleep and "would write equations on his tongue" [11, p. 36]. When people tell me this, I assure them that I have been a witness to it. I saw it happen on a stage in Berkeley.

Krishnaswami Alladi tells in his review of the play in *The Hindu* [2] that when he and George Andrews arrived in San Francisco for meetings of the American Mathematical Society that year and found that *Partition* was playing, they decided they had to see it, but seats were scarce. (They would have been, of course. It was a play about mathematics!) Miraculously, two seats were available for a Saturday night performance. Alladi credits this good fortune to the intervention of the goddess Namagiri! A wide-ranging review of the play by Kenneth A. Ribet appeared in the *Notices of the American Mathematical Society* [16]. The story of Ramanujan had reached the musical world even earlier:

an opera, *Ramanujan*, by Sandeep Bhagwati, was premiered in Munich, April 21, 1998 [5].

More recently, a multimedia presentation by Complicité (earlier known as the Théâtre de Complicité), *A Disappearing Number*, by Simon McBurney (music by Nitin Sawhney), opened in Plymouth, England, in 2007, and later played at various theater festivals in Holland, Germany, and Austria, as well as at the National Theatre in London. Eventually, it was broadcast to cinemas worldwide via *National Theatre Live*. The plot ran over two time periods, a historical section in Cambridge when Ramanujan and Hardy met, as well as a contemporary and fictional account of a mathematician and her husband, paralleling in a general way the earlier story. The playwright was inspired by his reading Hardy's *A Mathematician's Apology* [10]. The play enjoyed considerable success, receiving several theater awards, including the prestigious 2008 Laurence Olivier Award for Best Play. Unfortunately, it did not play widely in the United States (only in 2010 in Ann Arbor, Michigan, and at the Lincoln Center Festival in New York), but it did appear in many venues abroad, in Milan, Barcelona, Paris, Sydney, and, not surprisingly, in Mumbai and Hyderabad, where it played during the International Congress of Mathematicians in the summer of 2010.

Ramanujan left a legacy of provocative formulas that have prompted generations of mathematicians to try to understand his conjectures and eventually provide proofs. Three American mathematicians have been at the forefront in continuing these investigations: George Andrews of the Pennsylvania State University, Richard Askey at University of Wisconsin–Madison, and Bruce Berndt at University of Illinois at Urbana–Champaign. Berndt has described a pilgrimage in India in [6] and published two volumes on Ramanujan's notebooks [7]. The first is a touching account Berndt gives of a trip to India to visit Ramanujan's home and schools and to meet members of his family. An earlier article by Andrews appeared in the *American Mathematical Monthly* in 1979 [3] and included accounts of some of Ramanujan's later mathematical work to complement those provided earlier by Hardy, who had no access to the "lost notebooks" that Andrews discovered in some papers of another mathematician at Trinity College, Cambridge, in 1976: roughly 100 pages of densely packed formulas written in Ramanujan's hand. Periodically, there appear summaries of progress on proving the conjectures in the notebooks. And the work goes on.

FIGURE 1. Bronze bust of Ramanujan by Paul T. Granlund.
Courtesy of Leonard F. Klosinski

With so much having been written about Ramanujan over the many years since his death, one might wonder why someone as relatively unfamiliar with the subject as I am would attempt to add to the literature about Ramanujan, other than for the pleasure I have had in reviewing the elegant pieces written by Hardy and others, including Hardy's short summary of his own life, his *A Mathematician's Apology*. One should never pass up an opportunity to read that small book again, something that Atle Selberg referred to as "a great piece of literature" in a conversation we had in 1999 [1, p. 266].

And now to the curious event that prompted this note. In August 2011, I received an e-mail from the noted geometer and Escher expert, Doris Schattschneider, informing me that a bronze bust of Ramanujan was coming up for auction in Philadelphia at Freeman's (the oldest active American auction house). She wondered whether I knew of anyone who would be interested. The lot was hidden away at the end of a catalogue of American paintings, drawings, and sculpture, not likely to attract wide attention in the mathematical world. Well, the answer was obvious. I bid on it, and the bust is now in my office. But the story is complicated. Kanigel mentioned in his biography that Ramanujan's wife Janika had raised the question with Richard Askey of why, after her being promised that there would be a statue of Ramanujan in his home town, Kumbakonam (he was born nearby in Erode), she was still waiting for a statue. Berndt calls Kumbakonam a town, though it has a population of roughly 150,000. Askey responded by commissioning the American sculptor Paul T. Granlund to produce a bust using the passport photograph taken when Ramanujan left for England. Granlund was a prolific artist, perhaps best known for his sculpture of Charles Lindbergh at Le Bourget outside Paris, the airfield where Lindbergh landed after his 1927 solo flight over the Atlantic. Other casts of this sculpture can be seen on the grounds of the state capitol in St. Paul, Minnesota, and at the terminal at Lindbergh Field in San Diego. I contacted Askey to find out where my copy of the sculpture might have come from, and in his response he outlined the history of the work [5]. There were ten copies cast in 1983 (plus an artist's proof). One was given to Ramanujan's wife, Janika, and is now in the Ramanujan Institute of Mathematical Sciences in Madras (Chennai). Four others are in India, at research institutes in Delhi, Poona, Bangalore, and Mumbai. One that was originally acquired by S. Chandrasekhar is in London at the headquarters of the Royal Society, another in a building near the Isaac Newton Institute at Cambridge. The other three are in the United States, one owned by Askey, one by George Andrews, and the one in my office. (The artist's proof is at Gustavus Adolphus College in Minnesota, where Granlund was artist-in-residence.) My copy is almost certainly the one that was once owned by James Vaughn, the Texas oilman and philanthropist who was for many years generous to various mathematical organizations. He died in 2007.

Endnote: All right. How much did I pay for it? Someone is sure to ask. The auction house estimate in the sale catalogue was already low—US$1,000–$1,500—and I was ready to bid far more than that. But when the lot came up, the auctioneer at the sale announced that the opening bid for that lot would have to be $500. I bid that, and there were no further bids. Of course, I paid considerably more than that since there was a 25% premium on the hammer price as well as an even more alarming bill for putting it in a crate and shipping it to California. Nevertheless, it was something of a steal.

References

[1] Donald J. Albers and Gerald L. Alexanderson, eds. *Fascinating Mathematical People/Interviews and Memoirs*, Princeton University Press, Princeton, NJ, 2011. MR2866911.

[2] Krishnaswami Alladi. *Partition: A Play on Ramanujan, The Hindu* (Chennai-Madras), May 26, 2003.

[3] George E. Andrews. "An Introduction to Ramanujan's "Lost" Notebook." *Amer. Math. Monthly* **86** (1979), no. 2, 89–108, DOI 10.2307/2321943. MR520571.

[4] George E. Andrews. *The Theory of Partitions* (*Encyclopedia of Mathematics and Its Applications*, Vol. 2), Addison-Wesley, Reading, MA, 1976. MR0557013 (58# 27738).

[5] Richard Askey. *Personal Correspondence*, September 10, 2011. (This included the transcript of a talk given to the Madison Literary Club: "Romance in Mathematics—The Case of S. Ramanujan.")

[6] Bruce C. Berndt. "A Pilgrimage." *Math. Intelligencer* **8** (1986), no. 1, 25–30, DOI 10.1007/BF03023916. MR823217.

[7] Bruce C. Berndt. *Ramanujan's Notebooks*, 5 parts, Springer, New York, 1985–1997.

[8] Godfrey Harold Hardy, P. V. Seshu Aiyar, and B. M. Wilson, eds. *Collected Papers of Srinivasa Ramanujan*, Cambridge University Press, Cambridge, London, and New York, 1927. (Obituary reprinted from *Proc. London Math. Soc.* (2), xix (1921), pp. xl–lviii.) Reissued by the American Mathematical Society (Chelsea), Providence, RI, 2000.

[9] Godfrey Harold Hardy. "The Indian Mathematician Ramanujan." *Amer. Math. Monthly* **44** (March 1937), no. 3, 127–155. Reprinted in G. H. Hardy, *Ramanujan: Twelve Lectures on Subjects Suggested by His Life and Work*, Cambridge University Press, Cambridge, 1940. Reissued by the American Mathematical Society (Chelsea), Providence, RI, 1999. MR1523880; MR0004860; MR0106147 (21:4881).

[10] Godfrey Harold Hardy. *A Mathematician's Apology,* Cambridge University Press, Cambridge, 1940.

[11] Robert Kanigel. *The Man Who Knew Infinity: A Life of the Genius Ramanujan*, C. Scribner's, New York, 1991. MR1113890 92e:01063.

[12] Gary F. Kurutz. *Introduction*, Zamorano Select, Zamorano Club, Los Angeles, 2010, p. 6.

[13] David Leavitt. *The Indian Clerk: A Novel*, Bloomsbury, New York, 2007.

[14] David Leavitt. *The Man Who Knew Too Much: Alan Turing and the Invention of the Computer,* W. W. Norton, New York, 2007.

[15] Srinivasa Ramanujan. "Letter to G. H. Hardy, 16 January 1913." in *The G. H. Hardy Reader*, by D. J. Albers, G. L. Alexanderson, William Dunham, eds., Cambridge University Press, Cambridge, and Mathematical Association of America, Washington, DC, 2016.

[16] Kenneth A. Ribet. "Partition." *Notices Amer. Math. Soc.* **50** (December 2003), no. 11, 1407–1408.

Creating Symmetric Fractals

Larry Riddle

Fractals such as the Sierpinski triangle, the Koch curve, and the Heighway dragon, shown in Figure 1, are constructed using simple rules, yet they exhibit beautifully intricate and complex patterns.

All three fractals possess *self-similarity*—that is, the fractal is composed of smaller copies of itself. But only the first two fractals have symmetries. In this article, we show how to use group theory, which is often used to describe the symmetries of objects, to create symmetric fractals.

Iterated Function Systems

Fractals such as those in Figure 1 can be constructed using sets of functions called *iterated function systems* (IFSs). The functions in an IFS have a scaling factor less than 1, a rotation, and a translation, which makes them contractive affine transformations.

For instance, the IFS for the Heighway dragon is the set of functions $H = \{h_1, h_2\}$ where h_1 is a scaling by $r = \frac{1}{\sqrt{2}}$ and a counterclockwise rotation by 45° around the origin, while h_2 is also a scaling by r but with a counterclockwise rotation by 135° and a horizontal translation by 1. The dragon is the unique set A, called the *attractor* of the IFS, that satisfies $A = h_1(A) \cup h_2(A)$. In Figure 1, $h_1(A)$ is dark gray and $h_2(A)$ is light gray (in color Figure 1c, red and blue, respectively). Both subsets are copies of the Heighway dragon scaled by a factor of r.

Bringing in Abstract Algebra

In *Symmetry in Chaos: A Search for Pattern in Mathematics, Art and Nature* (1992, Oxford University Press), Michael Field and Martin Golubitsky showed how to generate a symmetric fractal from a single affine

FIGURE 1. Clockwise from far left, the Sierpinski triangle, the Koch curve, and the Heighway dragon. See also color images.

transformation and a *cyclic group* Z_n or a *dihedral group* D_n—groups that students encounter in an abstract algebra course. As we shall see, their method applies equally well when applied to an iterated function system.

First, we need some basic facts about Z_n and D_n. The group Z_n consists of the rotational symmetries of a regular n-sided polygon. We will let the n elements of Z_n correspond to counterclockwise rotations about the origin (which corresponds to the center of the polygon) through angles that are integer multiples of $360°/n$.

The group D_n consists of all $2n$ symmetries of a regular n-sided polygon. We take these to be the n rotations in Z_n and reflections about n lines through the origin that meet in angles that are integer multiples of $180°/n$. Figure 2 shows the lines of symmetry for an equilateral triangle and a square. The group D_2 is an exception since there is no regular polygon with two sides. It is known as the *Klein four-group*, and the four elements are the identity, a vertical reflection, a horizontal reflection, and a 180° rotation.

Given any IFS, we can form a new one by composing each element of the group (Z_n or D_n) with each function in the IFS. For instance, let $Z_2 = \{e, g\}$, where e is the identity and g is the 180° rotation about

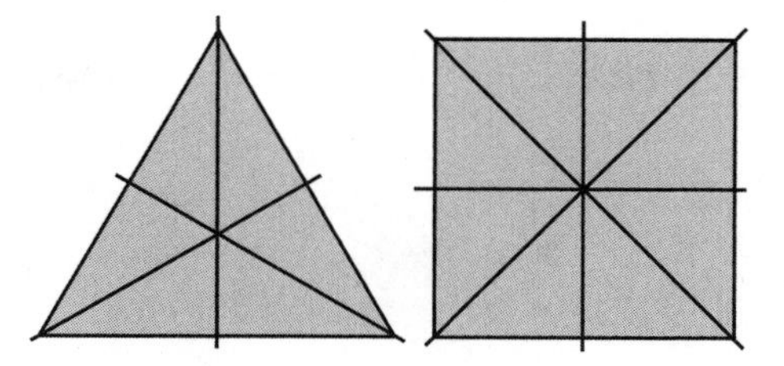

FIGURE 2. The groups D_3 and D_4 consist of rotations about the centers of the equilateral triangle and the square, respectively, and reflections about the lines of symmetry. See also color image.

the origin. Composing elements of Z_2 with elements in the IFS for the Heighway dragon, H, yields the IFS

$$\{f_1 = e \circ h_1 = h_1, f_2 = e \circ h_2 = h_2, f_3 = g \circ h_1, f_4 = g \circ h_2\}.$$

All four functions in this IFS have the same scaling factor, $\frac{1}{\sqrt{2}}$, because the only change—if any—is a 180° rotation. So, in addition to the original counterclockwise rotations of 45° and 135° from f_1 and f_2, the elements f_3 and f_4 have counterclockwise rotations of 225° and 315°. Also, f_2 and f_4 include horizontal translations by 1 and −1, respectively. The unique attractor for this new IFS, B, which we call the Z_2 Heighway dragon, satisfies $B = f_1(B) \cup f_2(B) \cup f_3(B) \cup f_4(B)$ (Figure 3).

Rotating B by 180° is the same as applying g from Z_2 to the set B. Because $g \circ g = e$ is the identity and $g \circ e = g$, we have

$$\begin{aligned} g(B) &= g(f_1(B) \cup f_2(B) \cup f_3(B) \cup f_4(B)) \\ &= g \circ h_1(B) \cup g \circ h_2(B) \cup g \circ g \circ h_1(B) \cup g \circ g \circ h_2(B) \\ &= g \circ h_1(B) \cup g \circ h_2(B) \cup e \circ h_1(B) \cup e \circ h_2(B) \\ &= f_3(B) \cup f_4(B) \cup f_1(B) \cup f_2(B) = B. \end{aligned}$$

This shows that the Z_2 Heighway dragon has twofold (or 180°) rotational symmetry.

In Figure 3, we see only three scaled versions of the attractor, but four sets are in the union. Because f_1 and f_3 have the same scaling factor, and they rotate B by 45° and 225° counterclockwise, respectively, and B has 180° rotational symmetry, then $f_1(B) = f_3(B)$. This is the black set (red in color Figure 3).

FIGURE 3. The Z_2 Heighway dragon. See also color image.

Coloring a Fractal with Pixel Counting

One of the methods for generating the attractor of an iterated function system on a computer is to use a random algorithm known as the *chaos game*. First, we choose an initial point x_0 in the attractor. Then we choose a function f from the IFS at random (with a certain probability), and then we plot $x_1 = f(x_0)$, which lies in the attractor.

Next, we choose a function from the IFS at random again, apply it to x_1, and obtain another point in the attractor, x_2. Repeat this process of randomly choosing a function, plugging in the previous point, then plotting the new point—millions of times. The sequence of points fills out the attractor.

In fact, it is not necessary to start with a point in the attractor. Any point suffices as long as we wait long enough during the iterative process to start plotting the points. This allows time for the sequence of points to converge toward the attractor.

We can color a point based on which function was used to compute it. This is how we colored the fractals in Figure 1. For example, the

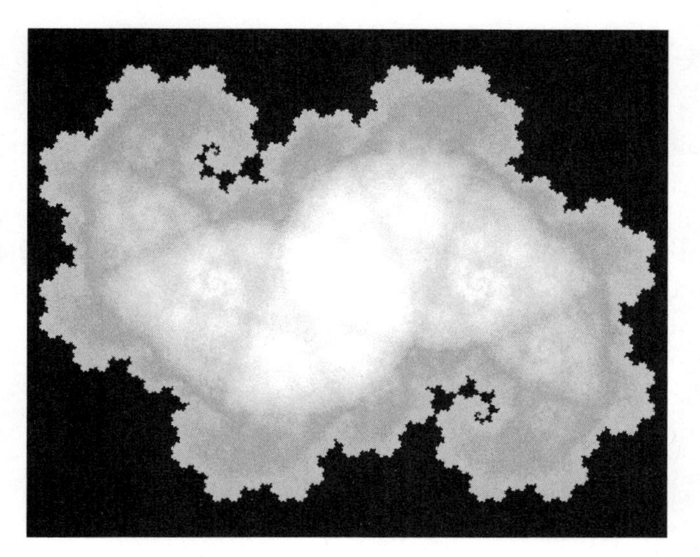

FIGURE 4. The Z_2 Heighway dragon colored by pixel counting. See also color image.

points in the Heighway dragon computed using h_1 are dark gray, and the points computed using h_2 are light gray (red and blue, respectively, in color Figure 1c.)

Of course, even the highest resolution computer screen has only finitely many points, or *pixels*. A single pixel can represent infinitely many points in the attractor. An alternative coloring scheme is to color a pixel based on how many points in the random sequence land in that pixel.

Figure 4 shows an image of the Z_2 Heighway dragon colored in this way. The color gradient varies from darker gray for low pixel counts through various shades of gray to pale gray as the pixel counts increase (in color Figure 4, the color gradient varies from gold for low pixel counts through various shades of dark orange to pale orange). Some experimentation is needed to find a good color gradient and the corresponding pixel counts. Factors to take into account include the size of the final image, the resolution of the computer screen, the number of points to be plotted, and the probabilities assigned to the functions in the IFS. But an appropriate choice helps to vividly illustrate the symmetry of the attractor.

Dragons and More

The Heighway dragon is just one example of a larger class of dragon fractals. Another example is the golden dragon curve, so named because its fractal dimension is the golden ratio. Figure 5 shows the golden dragon (formed from an IFS consisting of two functions) and the attractor obtained by composing the cyclic group Z_2 with this IFS. We used the same pixel coloring as for the Z_2 Heighway dragon. This attractor has 180° rotational symmetry.

Figure 6 shows the attractor obtained by composing D_2 with the IFS for the Koch curve, colored from dark gray to light gray using pixel counting (from red to violet in color Figure 6). The group D_2 gives the

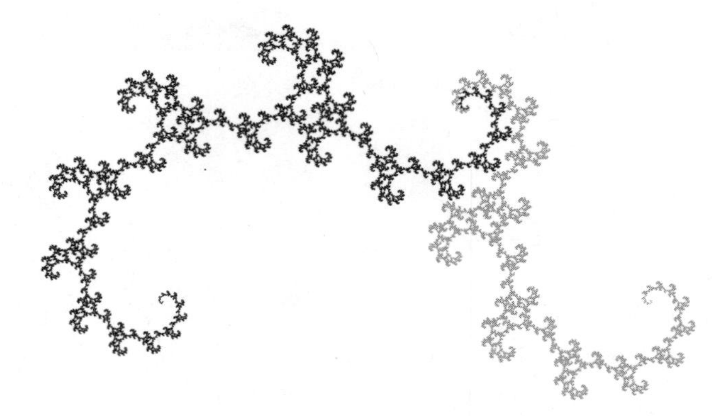

FIGURE 5. Golden dragon (top) and Z_2 golden dragon (bottom). See also color images.

FIGURE 6. The D_2 Koch curve. See also color image.

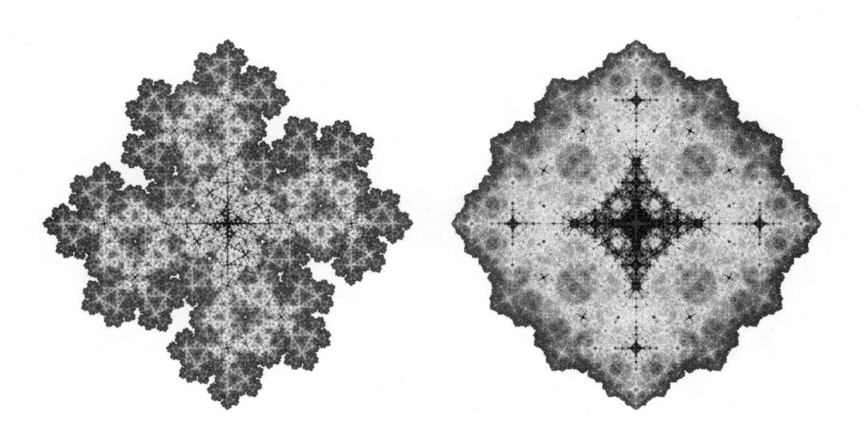

FIGURE 7. The Z_4 Koch curve (left) and the D_4 Koch curve (right). See also color images.

fractal 180° rotational symmetry and reflective symmetry across the horizontal and vertical lines through its center.

Figure 7 shows the fractals we obtain when we compose Z_4 and D_4 with the IFS for the Koch curve. The former has 16 functions that produce an attractor with fourfold rotational symmetry. The latter has 32 functions with an attractor displaying the D_4 symmetry of a square.

Finally, Figures 8 and 9 illustrate that if the original attractor already has some symmetry, then the attractor constructed by composing with

FIGURE 8. The Z_3 Sierpinski triangle. See also color image.

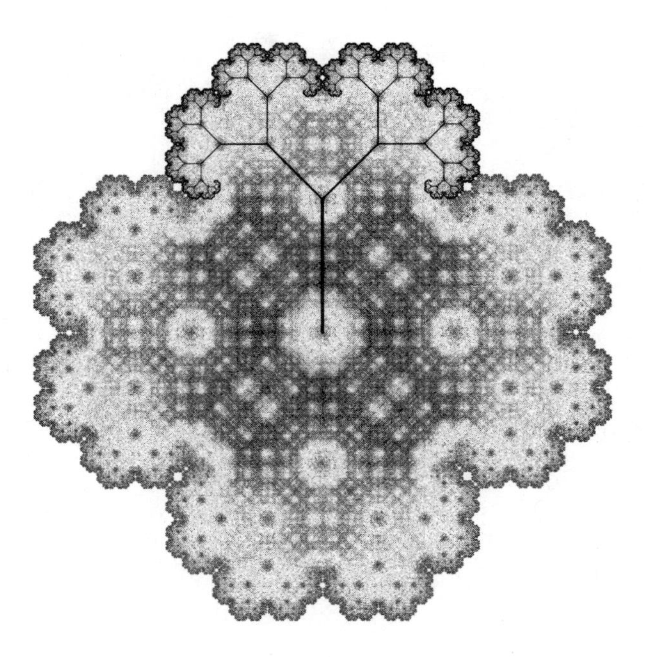

FIGURE 9. The Z_4 symmetric binary tree. See also color image.

Z_n or D_n may produce additional symmetries beyond those guaranteed from the group used to build it.

Figure 8 comes from composing Z_3 with the Sierpinski triangle IFS. It has D_6 symmetry because the attractor is a filled-in hexagon. The initial IFS used for Figure 9 generated the self-contacting symmetric binary tree with angle 45°, which is the black tree superimposed on the fractal. We composed Z_4 with the IFS to form an IFS whose attractor is shown. It has D_4 symmetry.

Further Reading

For more on symmetric fractals, see "Symmetric Fractals" (chapter 7) in Field and Golubitsky's *Symmetry in Chaos*; an updated second edition of their book came out in 2009 (SIAM).

More details can also be found at the author's website, ecademy.agnesscott.edu/~lriddle/ifs/ifs.htm.

All fractal images shown here were drawn with the Windows program IFS Construction Kit, available at ecademy.agnesscott.edu/~lriddle/ifskit/.

Projective Geometry in the Moon Tilt Illusion

MARC FRANTZ

This article is about an illusion of nature called the "moon tilt illusion" and ways in which it illuminates, and is illuminated by, perspective drawing and projective geometry. I focus on the illusion for a gibbous moon—that is, the case when most of the illuminated half is visible from the earth. (For more general discussions that include the crescent moon case, see [4, 6, 7].) I begin by reviewing some basic properties of perspective. In Figure 1, an observer stands on a horizontal ground plane, with one eye closed and the other eye at the point O called the viewpoint or center of projection. With her viewing eye, she looks at a vertical picture plane, onto which is projected an image of the colored part of the ground plane, which lies on the far side of the picture plane. The image is created as follows. Given a point such as A', its perspective image in the picture plane is the point A such that A', A, and O are collinear. Assigning points and their images corresponding colors, the image that forms is a familiar one, with a horizon line h at eye level, and the images of the parallel sides of the light road converging on the horizon at a vanishing point T. The line of sight OT is parallel to the sides of the road, lines $A'C'$ and $B'D'$, which are perpendicular to the picture plane. The road and grass parts of Figure 1 illustrate the basic principles of linear perspective as conceived by artists of the Renaissance. Intuitively, the idea is that when the observer at O looks at the picture plane, she sees the same thing she would see when looking at the ground plane, since the same colors come from the same directions.

Observe that there is no point in the ground plane corresponding to the point T, since the line of sight OT does not meet the ground, being parallel to it. In projective geometry, this situation is addressed by defining an abstract point at infinity T' belonging to OT and all lines parallel

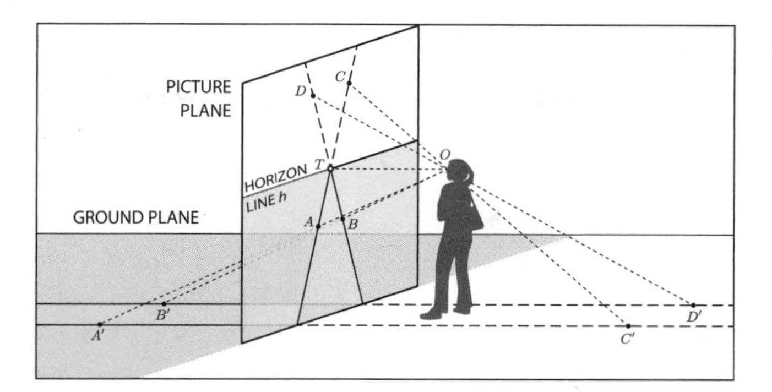

FIGURE 1. One-point projection of the ground plane onto the picture plane (Credit: Marc Frantz).

to it, such as $A'C'$ and $B'D'$. The point T is thus the perspective image of T'. The line OT with the point T' included is called an extended line. Similarly, there is an abstract line at infinity h' in the ground plane, whose image is the horizon line h. The line h' belongs to any plane parallel to the ground plane. The ground plane with the line h' included is called an extended plane. One way in which projective geometry dramatically departs from perspective drawing is the custom of projecting every point of one extended plane, such as the ground plane, onto another, such as the picture plane. Thus, for example, the point C' in the ground plane projects to the point C in the picture plane such that C', C, and O are collinear. The result is that points like C' and D', which lie behind the viewer, project above the horizon line, causing part of the ground to appear in the sky! Another consequence is that every infinite extended line in the ground plane such as $A'C'$ has as its image an infinite extended line in the picture plane, in this case AC. Although perspective drawings never explicitly include images of points behind the viewer, we will see that such projections can be very useful to the artist.

The Disk Tilt Illusion

Our starting point for understanding the moon tilt illusion is an illusion we call the "disk tilt illusion." We begin with Figure 2, a perspective drawing of three congruent, vertically aligned boxes (rectangular parallelepipeds). The horizontal edges of each box run either east–west or north–south.

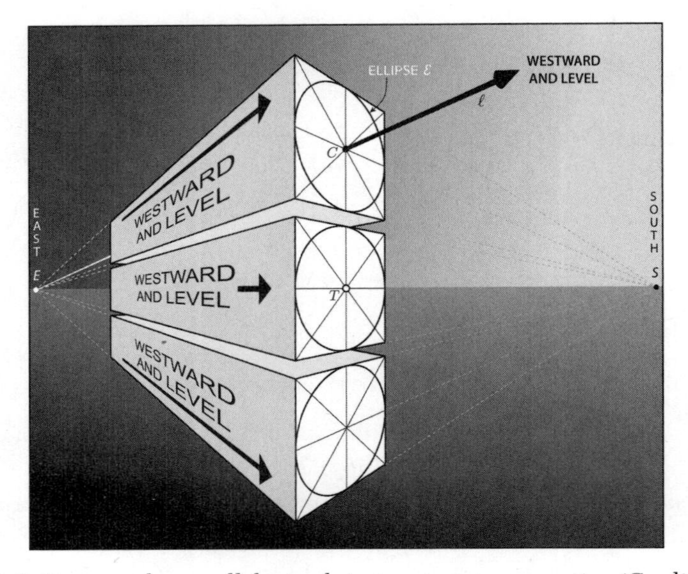

FIGURE 2. Rectangular parallelepipeds in two-point perspective (Credit: Marc Frantz).

When viewing Figure 2, we imagine ourselves as the observer in Figure 1, with one eye at a viewpoint O hovering in front of the the picture plane—the page—on a line perpendicular to the page at the viewing target T. The picture plane is again perpendicular to the horizontal ground plane. We think of the actual boxes as being on the far side of the picture plane, projected onto the picture plane via the center of projection O. In this case, the viewpoint is at a distance $|ET|$ $(= |TS|)$ from the point T. (This distance is rather close because of the limited size of the picture.) When looking directly at T, we are facing southeast, with the eastern horizon point E located 45° to the left of T and the southern horizon point S located 45° to the right of T. The points E and S are the vanishing points of the horizontal edges of the boxes. Technically, this means that E (respectively, S) is the perspective image of the point at infinity common to all east–west lines (respectively, to all north–south lines).

As indicated in Figure 2, the long edges of the boxes run "westward and level," meaning that they are aligned east–west and parallel to the ground plane. Their perspective images, on the other hand, fan radially outward from the eastward vanishing point E. The same is true of the arrows painted on the sides of the boxes, and the arrow protruding from the end of one box.. That is, the actual arrows all point westward and level, even though their images do not all point in the same direction. Thus if the sun is setting, located due west on the western horizon, all of the actual arrows point directly at the setting sun. Equivalently, the arrows point antiparallel to the sun's rays, which we idealize as parallel to one another.

Of course, I depend on the reader having looked at enough photographs and perspective drawings of buildings, fences, and so on, to readily accept that the four arrows in Figure 2 are the images of parallel arrows in the real world that point directly at the setting sun. Another perception I count on is the following: although the circles inscribed on the ends of the boxes have elliptical images whose major axes are inclined at various angles, each ellipse is the image of a circle in a vertical plane that faces westward and level (the normal to the plane of each circle points westward and level).

Most viewers would accept these relationships in Figure 2 because the boxes and their markings provide familiar clues as to what the image portrays. It is interesting to see what happens when most of this structure is removed. In Figure 3, we have removed everything but the

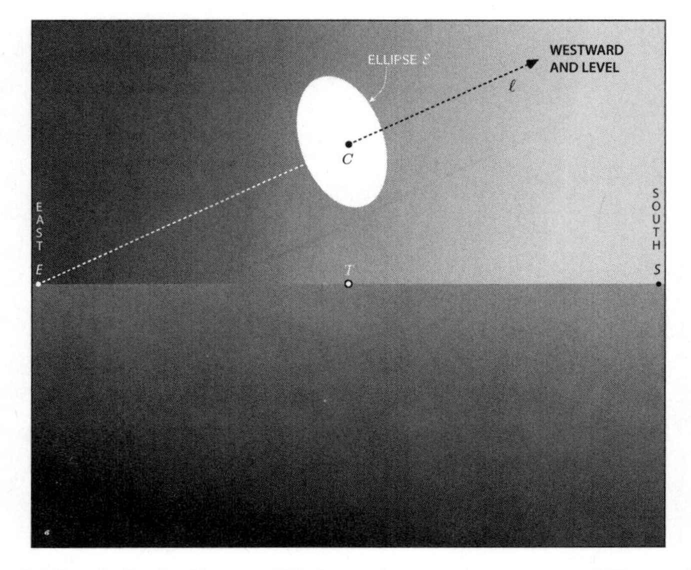

FIGURE 3. The disk tilt illusion. Without the extra structure of Figure 2, the disk no longer appears to lie in a vertical plane, but instead appears to face upward (Credit: Marc Frantz).

line l, the ellipse E, and the point C—the perspective image of the center of the circle whose image is E. It suddenly becomes more difficult to accept that E and its interior represent a circular disk in a vertical plane that faces westward and level. Rather, the disk appears to face at an upward angle—above the setting sun rather than toward it. This is the phenomenon we call the disk tilt illusion.

The Moon Tilt Illusion

The moon tilt illusion of a gibbous moon follows directly from the disk tilt illusion, in the following sense. We imagine the circle, whose image is E, as the boundary between light and dark—the "terminator"—of a sphere lit by the parallel east–west rays of the setting sun. The terminator is thus a great circle, in a westward-facing plane, of the sphere whose perspective image appears in Figure 4. In Figure 4, the perspective image of the sphere has an elliptical outline, which appears circular when viewed with one eye from the correct viewpoint (directly in front of T at a distance $|ET|$). Just as the disk of the terminator appears to face upward in Figure 3, the illuminated half of the moon (shown in

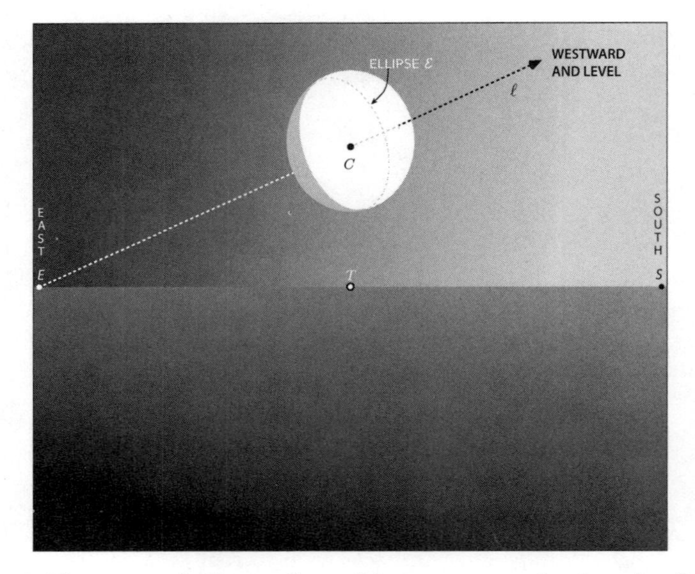

FIGURE 4. The moon tilt illusion for a gibbous moon can be thought of as a consequence of the disk tilt illusion, with the lunar terminator as the boundary of the disk (Credit: Marc Frantz).

white) appears to face upward in Figure 4. This illusion persists when observing the moon in nature. The perception of the illuminated side of the moon as facing above the setting sun instead of toward it is known as the moon tilt illusion.

To summarize, one way to understand the moon tilt illusion is by the progression from Figure 2 to Figure 3 to Figure 4. This progression illustrates the illusion as stemming from the lack of familiar structures (the boxes in Figure 2) that aid in perceiving the disk of the lunar terminator as being in a vertical, westward-facing plane, looking directly at the sun. With the structure removed in Figure 3, the same disk appears to face at an upward angle. For concreteness, we discussed the gibbous moon in the eastern sky with the sun setting in the west, but the discussion also applies to a gibbous moon in the western sky with the sun rising in the east.

The Journey of Sunlight

There is another way to perceive the moon in Figure 4 as directly facing the sun. In Figure 5, we look down on an observer with her viewing eye at the center of projection O, gazing at the moon through the picture plane, seen edge-on from above. Her eye is directly in front of the viewing target T, with the easterly vanishing point E at 45° to her left; thus she faces southeast toward the moon. The line l' through the

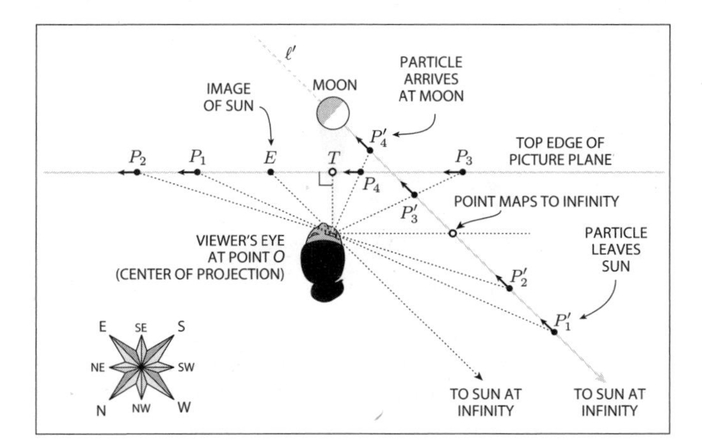

FIGURE 5. Journey of a particle from the sun to the moon (Credit: Marc Frantz).

center of the moon runs east–west and parallel to the ground plane. The sun, idealized as infinitely distant, is behind her on the western horizon. That is, we idealize the sun as the point at infinity on the parallel extended lines l' and OE, hence the point E is the perspective image of the setting sun. We follow the hypothetical journey of a moving point or particle, traveling along l' from the sun to the moon.

At the point P_1', the particle has left the sun, traveling east along l'. Its image P_1, being collinear with O and P_1', is to the left of the moon's image. At P_2', the particle has traveled closer to the actual moon, while its image P_2 is even further to the left. At some instant, the particle arrives at the white dot to the right of O, and its image is a point at infinity. Shortly after this, it arrives at P_3', and its image P_3 is to the *right* of the moon's image, moving to the left. Just before impact with the moon, the particle is at P_4', and its image P_4 is immediately to the right of the moon's image.

The observer's view of this sequence of events is sketched in Figure 6, where the images P_1, P_2, P_3, and P_4 are moving down and to the left along the extended line l, which is the perspective image of the extended line l'. When the viewing eye is at the correct viewpoint, the image of the moon appears symmetric about l. Thus the image l of the path of the particle appears normal to the outline of the image of the illuminated hemisphere. There is a sense in which the image of the lighted hemisphere faces "directly toward" the image E of the sun, because the only way to reach the image of the lighted side of the moon from E along l is via the sequence P_1, P_2, P_3, P_4 just described. In

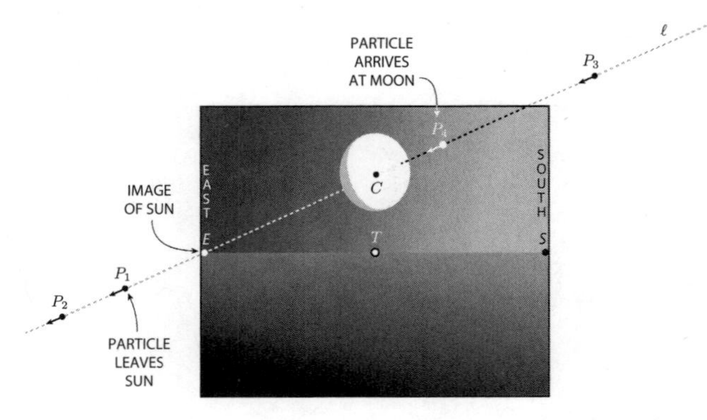

FIGURE 6. The particle's strange journey as seen by the viewer in Figure 5 (Credit: Marc Frantz).

this sense, the image is completely self-consistent, and there is nothing wrong with the tilt of the image of the lighted hemisphere.

A Shadow-Drawing Technique

This somewhat unusual resolution of the moon tilt illusion depends on a concept rarely used in basic perspective drawing, namely, the projection of unseen objects *behind* the viewer onto the picture plane. It is therefore interesting that this concept forms the basis of a convenient and effective drawing technique described by Rex Vicat Cole in his classic drawing manual [3, pp. 192–196]. The technique is illustrated in Figure 7, where the viewer of a painting has her viewing eye located at the correct viewpoint O, directly in front of the viewing target T. Once again, she faces southeast, with the evening sun in the west, over her right shoulder. This time, however, it is not quite sunset, so the sun is slightly above the western point on the horizon. The image P of the sun is therefore to the left and *below* the horizon. Cole refers to P as the "pseudo sun."

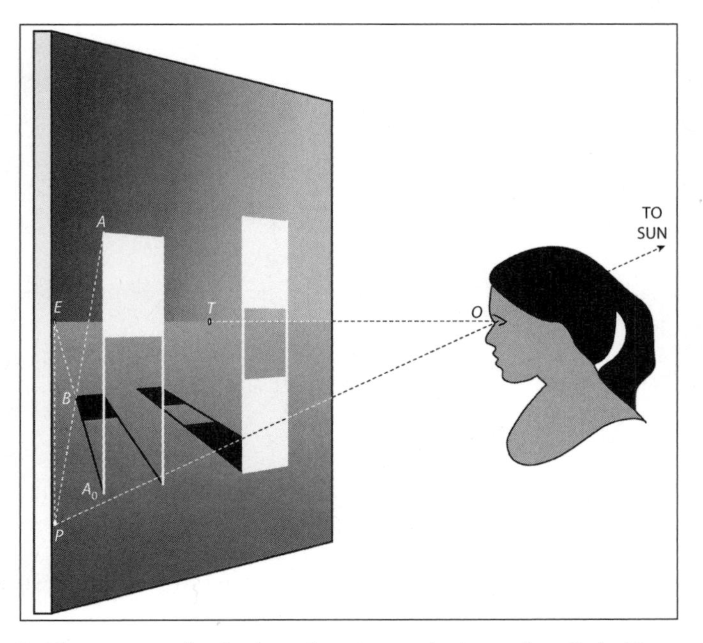

FIGURE 7. Illustration of a shadow-drawing technique after Cole [3, pp. 192–196] (Credit: Marc Frantz). See also color image.

FIGURE 8. Realistic evening shadows produced by Cole's technique (Credit: Marc Frantz).

To draw the shadow B of a point A, proceed as follows. Mark the point E on the horizon directly above P. Locate the image A_0 of the point on the ground plane directly below the preimage of A. The shadow B of A is then $EA_0 \cap PA$. (Since the sun is in the west, the shadow of the vertical line whose image is AA_0 is the east–west line with image EA_0, hence B lies on EA_0. But B must also lie on the light ray, or sunbeam, whose image is PA, hence $B = EA_0 \cap PA$.)

We tested Cole's technique in Figure 8 on three rectangles and four line segments. Even with such a simple configuration, the result is quite realistic and suggestive of late afternoon (or early morning) light.

Conclusion

At dusk the cock announces dawn;
At midnight, the bright sun.
—Poem from the Zenrin Kushu *[9, p. 117]*

The moon tilt illusion continues to be an object of interest, from the standpoints of light and optics [1, 2, 5], geometry and perspective [4, 6], human perception [7], and pop-science infotainment [8]. Here we presented the phenomenon as a reminder of the abstractions of projective geometry, and the marriage of opposites in nature. From the correct viewpoint in Figure 6—or in nature—we can see the symmetry line of the gibbous moon point directly to the image of the evening sun in the *east*, or the morning sun in the *west*, as the case may be. These abstractions lead to a practical drawing technique, with a pleasing artistic result.

Acknowledgment

The author is grateful to Annalisa Crannell for several helpful comments.

References

[1] Berry, M. V. "Nature's Optics and Our Understanding of Light." *Contemporary Physics*, **56** (1) (2015), 2–16. Available online at http://dx.doi.org/10.1080/00107514.2015.971625.
[2] Berry, M. V. "The Squint Moon and the Witch Ball." *New J. Phys.*, **17** (2015), 060201. Available online at http://dx.doi.org/10.1088/1367–2630/17/6/060201.
[3] Cole, R. V. *Perspective for Artists*, Dover, Mineola, NY, 1976.
[4] Glaeser, G., and Schott, K. "Geometric Considerations about Seemingly Wrong Tilt of Crescent Moon." *KoG (Croatian Soc. Geom. Graph.)*, **13**(13) (2009) 19–26. Available online at http://hrcak.srce.hr/index.php?show=clanak&id_clanak_jezik=73428.
[5] Minnaert, M., *Light and Colour in the Open Air*, Dover, New York, 1954.
[6] Myers-Beaghton, A. K., and Myers, A. L. "The Moon Tilt Illusion." *KoG (Croatian Soc. Geom. Graph.)*, **18**(18) (2015) 53–59. Available online at http://hrcak.srce.hr/index.php?show =clanak&id_clanak_ jezik=197562.
[7] Schölkopf, B. "The Moon Tilt Illusion." *Perception* **27**(10), (1998), 1229–1232.
[8] Stevens, M. "The Moon Terminator Illusion" (Vsauce YouTube video), published 6/7/2015. Accessed 12/11/2015 from https://www.youtube.com/watch?v=Y2gTSjoEExc.
[9] Watts, A. W. *The Way of Zen*, Vintage, New York, 1957.

Girih for Domes: Analysis of Three Iranian Domes

Mohammadhossein Kasraei, Yahya Nourian, and Mohammadjavad Mahdavinejad

Introduction

The rise and spread of Islamic art, during many consecutive years, has provided a great heritage of geometric patterns in art and architecture. A variety of geometric patterns is used by artists to embellish a wide range of works of art, including textiles, ceramics, metalwork, and architectural elements including Kar-bandi, Rasmi-bandi, Muqarnas, and Girih. The tiling design called Girih[1] in Persian (Figure 1), is one of the most significant and complicated techniques using geometric patterns to adorn art and architectural surfaces. The use of these geometric patterns has spread across the world, from west India to Southern Europe and Turkey, to the Arabian Peninsula. However, some believe that the mathematics and construction of them originated in Iran (Necipoğlu and Al-Asad 1995).

Although these tessellations consist of some restricted and regular shapes (motifs), which have their own specific Persian names, such as Shamseh (star), Torange (the quadrilateral tile), Sormeh-Dan (the bow tie tile), Tab'l (the concave octagonal tile), etc., these shapes have generated various and distinct patterns in art and architecture (Ra'eesZadeh and Mofid 2011: 142).

Assessing many examples from the past thousand years reveals that the main feature of the Girih is the mathematical and geometrical principles used in drawing them (Figure 2). The diversity of pattern types is based on the variety of methods for drawing. It is useful for current designers to know how each pattern is drawn.

In recent years, many efforts have been made to analyze the process of drawing Girih. Some recent research reveals their geometric

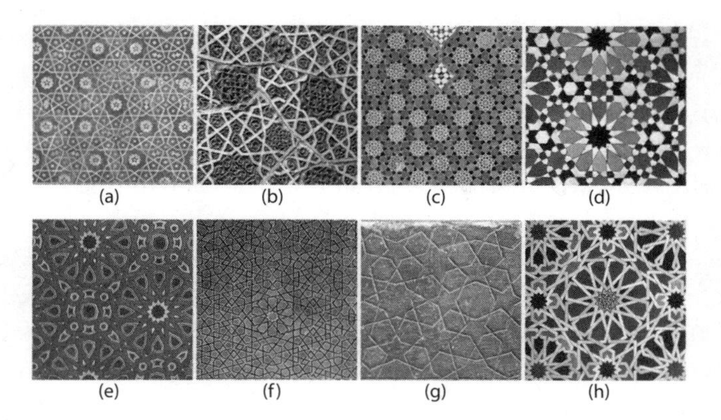

FIGURE 1. (a) Humayun's Tomb, New Delhi, India; (b) Shakhi Zindeh complex, Samarkand, Uzbekistan; (c) Masjid-i-Jami, Isfahan, Iran; (d) Bou Inaniya Madrasa, Fez, Morocco; (e) al-Nasir Mosque, Cairo, Egypt; (f) Karatay Madrasa, Konya, Turkey; (g) Altun Bogha Mosque, Aleppo, Syria; and (h) Alhambra, Granada, Spain (*Pattern in Islamic Art* 2015). See also color images.

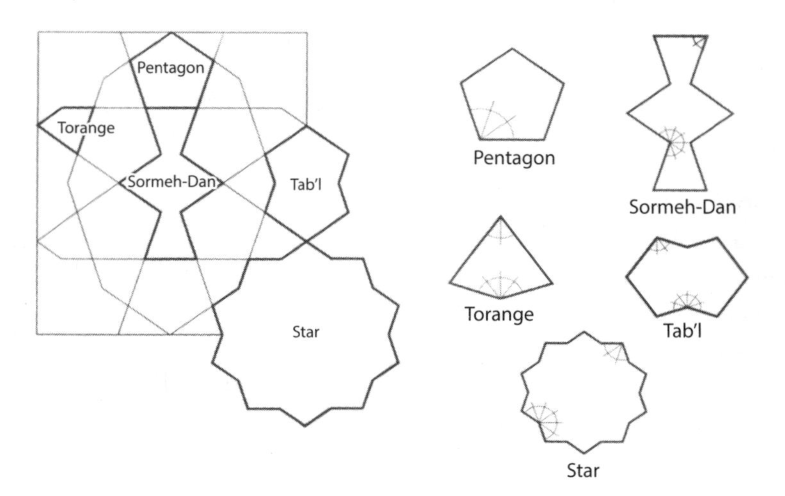

FIGURE 2. Some regular shapes (motifs) commonly used to construct Islamic patterns.

order and reinvents drawing instructions for Girih (Bonner 2003, Kaplan 2000, and El-Said and Parman 1976). In contrast, others pursue traditional methods and study documents such as scrolls and booklets, enabling active professional artisans to redefine original methods (Sarhangi 2012, Bodner 2012, and Ra'eesZadeh and Mofid 2011). It seems that achieving a comprehensive theory covering the variety of Girih types will require more effort in the future. Previous research might be summarized in two major drawing approaches: the Radial Grid approach and the Polygons in Contact approach.

The Radial Girih approach is a traditional method illustrated in many ancient documents such as "On Interlocking Similar or Corresponding Figures" (Fi tadakhul al-ashkal al-mutashabihat aw al-mutawafiqa), written by Abu Ishaq Ibn Abdullah Koubnani, which was attached to "A Book on Those Geometric Constructions Which Are Necessary for a Craftsman"(Kitāb fī mā yaḥtāj ilayh al-ṣāniʿ min al-aʿmāl al-handasia), written by Abu al-Wafa' al-Būzjānī (June 10, 940–July 15, 998). The translation of this book is represented in *Applied Geometry* (Jazbi 1997). This method was also taught by some Iranian professional artisans, who inherited their profession from their ancestors dating back several centuries. These include Hussein Lorzadeh (Ra'eesZadeh and Mofid 2011), Mahmoud Maheronnaqsh (1984), Asghar Shaarbaf (1385), etc. In this approach, drawing the Girih begins by placing the centers of stars a specific distance from each other. Depending on the number of star polygons, specific numbers of rays emanate from the centers of stars. The intersections of these rays provide interstitial space, which is filled by different methods and creates various designs. In some medieval samples, a pattern covers a simple large surface by repeating and replicating a basic part of patterns called the repeat unit.[2] In these simplified cases, the boundary of the repeat unit is constructed according to geometrical principles and the measurement of the center of each star in the corners of the repeat unit's boundary (Sarhangi 2012: 167). This basic unit consists of fundamental information about the Girih, such as the number of points of star polygons, the placement, and the distance between the centers of the star polygons. Many of these repeat units have been found in specific resources, such as preserved Topkapi and Tashkent scrolls. Some researchers (Necipoğlu and Al-Asad 1995: 10–12) believe that the repeat unit drawings of these architectural scrolls most likely served as an aid to memorization for architects and master builders.

The Polygons in Contact approach is the another method, articulated by E. H. Hankin in the early part of the twentieth century (Hankin 1925). In this method, a surface covered with a network consists of "polygons in contact" (Kaplan 2005: 177–180), which is an unbounded

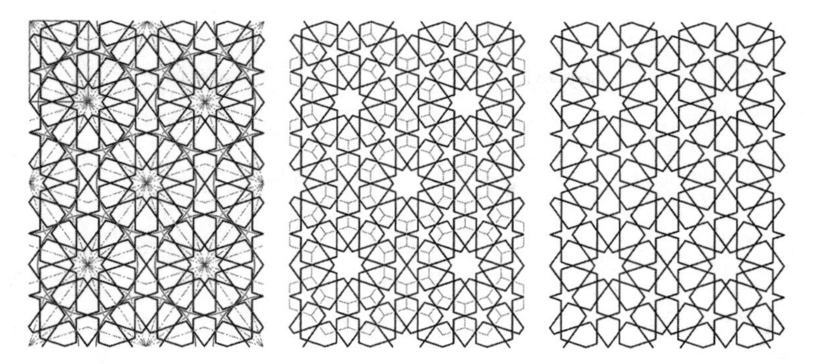

FIGURE 3. One specific Girih presented in Ra'eesZadeh and Mofid (2011). Drawn via the Radial Grid approach (left). Drawn via the Polygons in Contact approach (middle). (Drawn by the authors.) See also color images.

and infinite pattern, and it might be resized in order to fit a surface. Using the Polygons in Contact approach leads to regular and primitive patterns and, therefore, could not generate all the traditional types of Girih.

Investigating and comparing these two identified methods reveals that the Radial Grid approach is a general method that could generate a wide range of varied Girih, from traditional to new ones, and the Polygons in Contact approach is just a specific and restricted type that has constant and equal distance between the centers of stars (Figure 3).

This notion can be explained when we observe specific kinds of traditional Girih that are difficult or impossible to draw via the Polygons in Contact approach. For example, in traditional documents, there is a developed and elegant type of Girih called "Dast-Gardan"[3] in Persian (Figure 4).

The Dast-Gardan Girih is a developed kind of traditional Girih that consists of diverse types of stars fixed at different distances intentionally by the designer. The interstitial space among star polygons then is filled geometrically by basic regular shapes (motifs) such as Torange, Sormeh-Dan, Tab'l, etc. Recently, some researchers have attempted to reinvent a method to draw Dast-Gardan Girih. B. Lynn Bodner (2011) presented a method to draw a nine- and twelve-pointed star polygon design found in Tashkent scrolls. Although she did not mention anything about the Dast-Gardan concept and did not achieve a comprehensive method that covers the wide range of Dast-Gardan Girih, she outlines

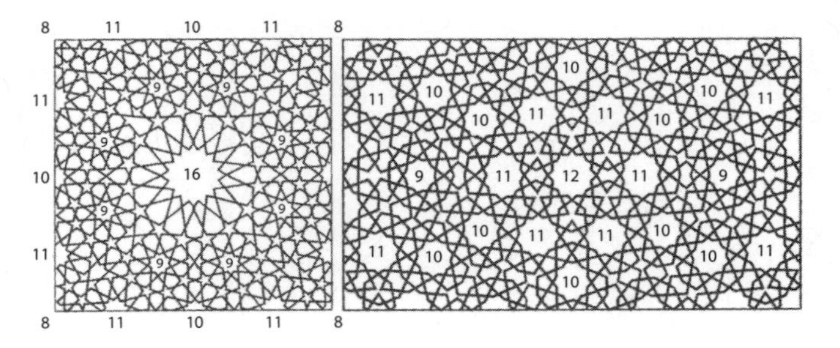

Figure 4. Two samples of Dast-Gardan patterns with various types of star polygons. The one on the left has 8-, 9-, 10-, 11-, and 16-pointed star polygons; the one on the right has 9-, 10-, 11-, and 12-pointed star polygons. Designed by Hussein Lorzadeh (Ra'eesZadeh and Mofid 2011: 178).

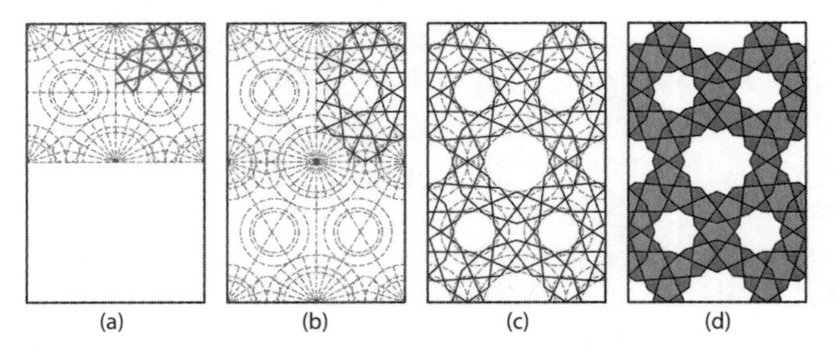

Figure 5. Redrawing a Dast-Gardan Girih designed by Hussein Lorzadeh (Ra'eesZadeh and Mofid 2011: 181). Fixing the center of circles with diverse diameters and drawing rays emanating from the centers (a); constructing the fundamental region (b); drawing the Girih unit (c); and drawing the complete Girih (d). See also color images.

a method to fill in the interstitial space with the irregularly shaped pentagonal star, hexagons, and arrowlike shapes. Generally, drawing Dast-Gardan Girih commences with fixing the center of diverse circles with various diameters. Then the circles divide to some congruent angles by creating rays that emanate from the center of circles. These rays intersect and then construct the fundamental region for the tiling, which consists of different types of star polygons. Figure 5 is an example of a step-by-step construction of a Dast-Gardan Girih designed by Hussein Lorzadeh (Ra'eesZadeh and Mofid 2011).

Unlike common Girih drawn with regard to an interactive basic unit and on an infinite surface, Dast-Gardan is constructed based on the form and shape of the surface. In other words, this type of Girih design is dependent on the surface, so the measurements and shape of the surface alter the process of its design. Dast-Gardan patterns have been utilized to impose geometric order on complicated surfaces and to have various options in order to draw a unique pattern. The role of Dast-Gardan becomes evident when a designer intends to draw a pattern on a curved surface.

Girih for Curved Surfaces

Professional Iranian artisans succeeded wonderfully in drawing patterns on curved surfaces and in making them fit the surfaces. A dome is one of the most significant curved surfaces and serves as a main characteristic of Islamic architecture; adorning domes is one of the challenging practices for Iranian artisans. A dome can be thought of as an arch that has been rotated around its central vertical axis and can therefore be categorized by its initial arch. However, many different types of arches are found in Islamic architecture throughout the medieval period. For example, a four-centered arch is a well-known arch utilized to shape many domes. The structure of a four-centered arch is achieved by drafting two arcs that rise steeply from each springing point[4] on a small radius and then turn into two arches with a wide radius and a much lower springing point. A dome made by a four-centered arch can be considered a smooth surface that has positive Gaussian curvature[5] ($K > 0$). As the Gaussian curvature is the product of the two principal curvatures ($K = k_1 k_2$), a dome-type surface at each point on a vertical section has unique Gaussian curvature (Figure 6).

On the other hand, according to historic documents, drawing Girih on the interior and exterior surface of a dome was a prominent challenge that many artisans faced (Hankin 1925: 23). The challenge is that patterns suitable to embed on the surface of domes do not follow the conventional methods of drawing and require a different approach (Figure 7).

No comprehensive study exists about drawing Girih on domes, but similar topics have been studied in recent research that attempts to draw patterns on a nonflattened surface. Kaplan and Salesin (2004) present a method to embed a specific pattern on different Euclidean and non-Euclidean surfaces. They state that "the structure of star

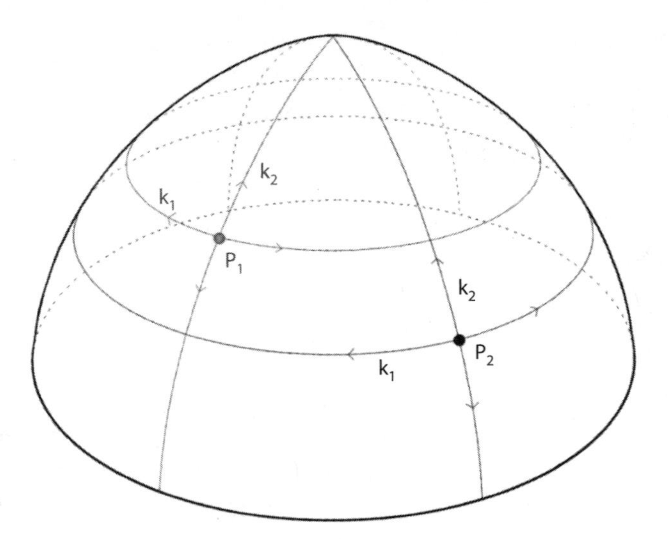

FIGURE 6. The Gaussian curvature ($K = k_1k_2$) at point 1 (P_1) is different from the curvature at point 2 (P_2).

FIGURE 7. Left, the interior dome of Yazd Jame' Mosque, middle, the exterior dome of the Saveh Jame' Mosque, and right, the interior dome of the Saveh Jame' Mosque. (Photographed by the authors.) See also color images.

patterns reflects the curvature of the space in which it is embedded" (Kaplan and Salesin 2004: 111). Consequently, the star polygons of a specific pattern change when placed on different surfaces with different curvatures. In other words, it may be concluded that the number of points of star polygons in Islamic patterns depends on the measurement of the "amount of space" (Kaplan and Salesin 2004: 111) around each point. Below we try to determine whether Kaplan and Salesin's conclusion could relate to Islamic patterns drawn by traditional artisans that decorated the surfaces of domes in Islamic architecture.

Analysis

Assessing the curved surfaces in Islamic architecture reveals that the Islamic patterns drawn on domes were usually designed by the Dast-Gardan method. This is not surprising, given the relationship between star polygons and the curvature of a surface. Based on the previous section, making a change in the number of star polygons applied to the pattern depends on the change of the dome curvature. So the pattern drawn on a dome should consist of diverse types of star polygons. This means that the patterns should be designed by the Dast-Gardan method. This can be confirmed by investigating the geometry of three scarce and special types of traditional domes in Iran and other Islamic regions that are covered by sophisticated and elegant Dast-Gardan Girih.

These three domes are the Yazd Jame' Mosque's dome (in Yazd, Iran) and the interior and exterior domes of the Saveh Jame' Mosque (in Saveh, Iran), which belong to two prominent and well-known buildings in Islamic architecture. Evidence suggests that the Saveh Jame' Mosque was constructed in 504 AH (in the year of the Hijra) during the Muhammad-ibn-Malek Shah Saljuqi period coinciding with the development of the city. Around the fourth or fifth century, the dome was constructed in the mosque's south front. The dome of the Saveh Jame' Mosque measures 14 m in diameter and 17 m high and is covered in Iranian Moarragh tessellations. The other building is the Yazd Jame' Mosque, which dates back to the Timurid dynasty, about 861 AH. The double-shelled dome of the Yazd Jame' Mosque is regarded as one of the masterpieces of Islamic architecture because of its geometry and patterns. These three domes have been selected because of the differences in their curvature so that the following discussion can represent the fact that the change in the curvature of the domes has a specific relation to the change in the Dast-Gardan Girih (Figure 8).

Investigating Dast-Gardan Girih drawn on domes shows that there is a significant relationship between the curvature of the surfaces of domes and the construction of Girih. For example, Figure 9 shows that the curvature of the surface of a dome changing from the spring point to the apex corresponds with the change in the number of points in star polygons in noticeable stages. Decreasing the curvature of a dome's surface leads to decreasing the number of points of star polygons.

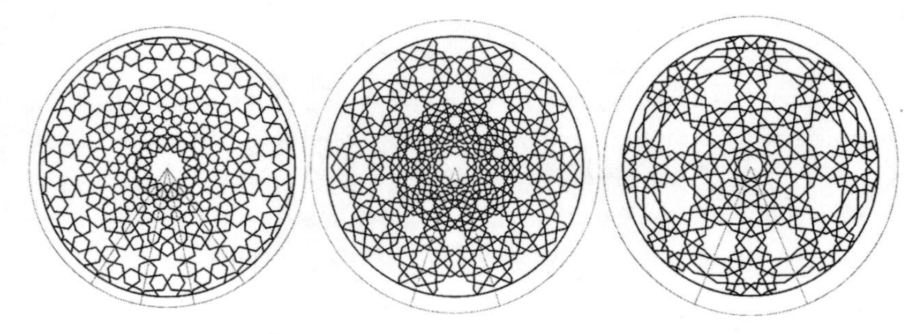

FIGURE 8. Left, Girih on the interior dome of the Yazd Jame' Mosque, middle, Girih on the exterior, dome of the Saveh Jame' Mosque, and right, Girih on the interior dome of the Saveh Jame' Mosque. (Drawn by the authors.)

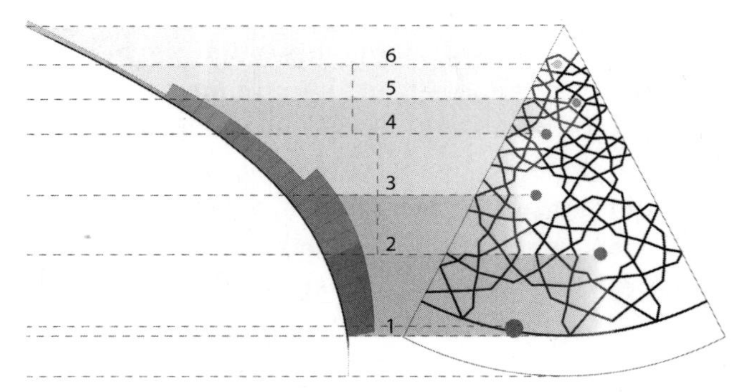

FIGURE 9. This diagram shows that the changing in curvature of a dome's surface corresponds with the change in number of points of star polygons. The diagram is a section of the exterior dome of the Saveh Jame' Mosque in Saveh, Iran. (Drawn by the authors.) See also color image.

Figure 10 shows the change in the Gaussian curvature of these three domes and the placement and number of points of each star polygon on the domes. Analyzing the Dast-Gardan patterns on these three domes reveals that the changing curvature is not the same in different cases. So the pattern that belongs to each dome has its own number of points of for star polygons. For example, the curvature of the exterior dome of the Saveh Jame' Mosque (Figure 10c) changes significantly in three stages from base to apex. In its interior dome, the curvature changes in two stages.

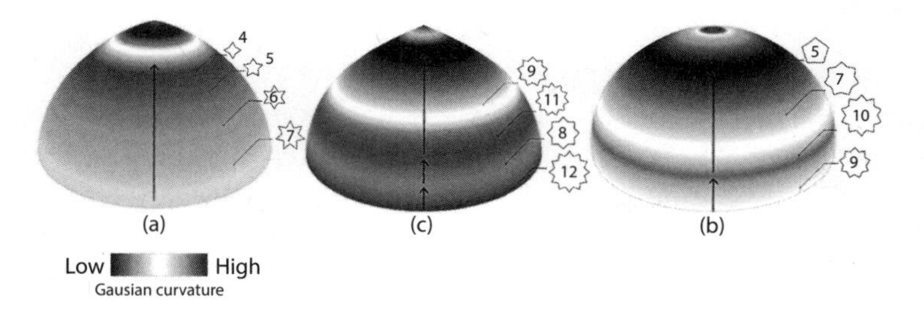

FIGURE 10. This diagram shows the Gaussian curvature and pattern of (a) the Yazd Jame' Mosque's dome, (b) the interior dome of the Saveh Jame' Mosque, and (c) the exterior dome of the Saveh Jame' Mosque. (The diagram was drawn using the Rhinoceros 5 program by the authors.) See also color images.

In contrast, the Yazd Jame' Mosque's dome is similar to a hemisphere, so the change in curvature is not great, and it changes in only one stage. Based on our assessment, changing curvature and the number of points of star polygons have a meaningful relationship. For instance, the pattern on the interior dome of the Yazd Jame' Mosque has four different star polygons—7, 6, 5, and 4—which relate to the changing curvature. Based on Figure 10a, the number of points of the star polygons increases steadily from 4 to 7. The patterns on the interior and exterior dome of the Saveh Jame' Mosque have a different trend. The changing curvature on the interior dome of the Saveh Jame' Mosque jumps in one stage, and its number of points for star polygons increases from 9 to 10 and then decreases to 7 and then 5. The changing curvature on the interior dome of the Saveh Jame' Mosque jumps in two stages. As seen in Figure 10c, first the number of points of star polygons goes down from 12 to 8, then it goes up to 11, and finally it decreases.

Conclusion

The variety of methods for drawing Islamic patterns leads us to one of the significant types, called Dast-Gardan in Persian. The importance of Dast-Gardan is evident when designers try to draw a pattern on a curved surface. A dome is one of the most significant curved surfaces, which has different Gaussian curvature at each point. The number of points for star polygons in Islamic patterns depends on the curvature of the surface; hence, the pattern embedded on a dome should consist of diverse types of star polygons. So the patterns on a dome should be a Dast-Gardan pattern in which its star polygons change according to the curvature of its surface. Analyzing the patterns of three traditional domes shows that there is a direct relationship between the changing curvature and the type and number of points for star polygons. As curvature increases throughout the surface, the pattern accommodates stars with larger numbers of points.

Notes

1. In this paper, we prefer to use "Girih" instead of "Islamic pattern," which is used in many new books and articles, because "Islamic pattern" is a vague term that consists of many geometric and nongeometric patterns.

2. A sample introducing a step-by-step procedure for generating a typical Girih by this approach is described in Sarhangi (2012).

3. *Dast-Gardan* is a Persian term that consists of two words; "Dast" is a noun which means "hand," and "Gardan" is a verb meaning "circulate." Iranian professional artisans used this term because Dast-Gardan patterns could be drawn freely according to the artisan's decisions.

4. The springing point is the point from which an arch and vaults spring or rise from their supports.

5 This definition is true because the curvature is the same way in two mutually perpendicular directions.

References

Bodner, B. L. 2011. "A Nine- and Twelve-Pointed Star Polygon Design of the Tashkent Scrolls." *Proceedings of Bridges 2011: Mathematics, Music, Art, Architecture, Culture*, Conference 2011, Reza Sarhangi and Carlo Séquin, eds., pp. 147–154. Phoenix: Tesselations Publishing. http://archive.bridgesmathart.org/2011/bridges2011–147.pdf.

Bodner, B. L. 2012. "The Topkapı Scroll's Thirteen-Pointed Star Polygon Design." *Proceedings of Bridges 2012: Mathematics, Music, Art, Architecture, Culture*, Robert Bosch, Douglas McKenna, and Reza Sarhangi, eds., pp. 157–164. Phoenix: Tesselations Publishers. http://archive.bridgesmathart.org/2012/bridges2012–157.pdf.

Bonner, J. 2003. "Three Traditions of Self-Similarity in Fourteenth and Fifteenth Century Islamic Geometric Ornament." *Meeting Alhambra, ISAMA-BRIDGES Conference Proceedings*, Javier Barrallo et al., eds., pp. 1–12. Granada, Spain: University of Granada. http://archive.bridgesmathart.org/2003/bridges2003–1.pdf.

El-Said, I., and Parman, A. 1976. *Geometrical Concepts in Islamic Art*. London: World of Islam Festival Publ. Co.

Hankin, E. H. 1925. *The Drawing of Geometric Patterns in Saracenic Art,* vol. 15. Archaeological Survey of India, Calcutta, India, Government of India Central Publication Branch.

Jazbi, S. A. 1997. *Applied Geometry.* Soroush Press (in Farsi), Tehran, Iran.

Kaplan, C. S. 2000. "Computer Generated Islamic Star Patterns." *Bridges: Mathematical Connections in Art, Music, and Science*, Conference Proceedings 2000, Winfield, KS, Reza Sarhangi, ed., pp. 105–112. http://archive.bridgesmathart.org/2000/bridges2000–105.html.

Kaplan, C. S. 2005. "Islamic Star Patterns from Polygons in Contact." *Proceedings of Graphics Interface 2005*, pp. 177–185. Canadian Human-Computer Communications Society, Victoria, BC, Canada.

Kaplan, C. S., and D. H. Salesin. 2004. "Islamic Star Patterns in Absolute Geometry." *ACM Transactions on Graphics* (TOG) **23**(2): 97–119.

Maheronnaqsh, M. 1984. *Design and Execution in Persian Ceramics*. Tehran, Iran: Reza Abbasi Museum Press.

Necipoğlu, G., and Al-Asad, M. 1995. *The Topkapı Scroll: Geometry and Ornament in Islamic Architecture*. Topkapı Palace Museum Library MS H. 1956. Santa Monica, CA: Getty Center for the History of Art and the Humanities.

Pattern in Islamic Art. 2015. http://patterninislamicart.com/. Accessed February 20, 2015.

Ra'eesZadeh, M., and Mofid, H. 2011. Ehyā-ye Honar Hā-ye Az Yād Rafteh: Mabāni-ye Me'māri-ye Sonati Dar Iran Be Revāyat-e Ostād Hossein-e Lorzādeh (Revival of Forgotten Arts: Basics of Traditional Architecture in Iran as Narrated by Master Hossein-E Lorzadeh). Tehran, Iran: Enteshārāt-e Moulā (Moula Publications) (in Farsi).

Sarhangi, R. 2012. "Polyhedral Modularity in a Special Class of Decagram Based Interlocking Star Polygons." *Proceedings of Bridges 2012: Mathematics, Music, Art, Architecture, Culture*, Robert Bosch, Douglas McKenna, and Reza Sarhangi, eds., pp. 165–174. Phoenix: Tessellations Publishing. http://archive.bridgesmathart.org/2012/bridges2012–165.pdf.

Shaarbaf, A. 1385. *Girih and Karbandi*. Tehran, Iran: Sobhan Nour. (in Farsi).

Why Kids Should Use Their Fingers in Math Class

Jo Boaler and Lang Chen

A few weeks ago I (Jo Boaler) was working in my Stanford office when the silence of the room was interrupted by a phone call. A mother called me to report that her 5-year-old daughter had come home from school crying because her teacher had not allowed her to count on her fingers. This is not an isolated event—schools across the country regularly ban finger use in classrooms or communicate to students that they are babyish. This is despite a compelling and rather surprising branch of neuroscience that shows the importance of an area of our brain that "sees" fingers well beyond the time and age that people use their fingers to count.

In a study published last year, the researchers Ilaria Berteletti and James R. Booth analyzed a specific region of our brain that is dedicated to the perception and representation of fingers known as the *somatosensory finger area*. Remarkably, brain researchers know that we "see" a representation of our fingers in our brains, even when we do not use fingers in a calculation. The researchers found that when 8-to-13-year-olds were given complex subtraction problems, the somatosensory finger area lit up, even though the students did not use their fingers. This finger-representation area was, according to their study, also engaged to a greater extent with more complex problems that involved higher numbers and more manipulation. Other researchers have found that the better students' knowledge of their fingers was in the first grade, the higher they scored on number comparison and estimation in the second grade. Even university students' finger perception predicted their calculation scores. (Researchers assess whether children have a good awareness of their fingers by touching the finger of a student—without the student seeing which finger is touched—and asking them to identify which finger it is.)

Evidence from both behavioral and neuroscience studies shows that when people receive training on ways to perceive and represent their own fingers, they get better at doing so, which leads to higher mathematics achievement. The tasks we have developed for use in schools and homes (see below) are based on the training programs researchers use to improve finger-perception quality. Researchers found that when 6-year-olds improved the quality of their finger representation, they improved in arithmetic knowledge, particularly skills such as counting and number ordering. In fact, the quality of the 6-year-olds' finger representation was a better predictor of future performance on math tests than their scores on tests of cognitive processing.

Many teachers have been led to believe that finger use is useless and something to be abandoned as quickly as possible.

Neuroscientists often debate why finger knowledge predicts math achievement, but they clearly agree on one thing: That knowledge is critical. As Brian Butterworth, a leading researcher in this area, has written, if students aren't learning about numbers through thinking about their fingers, numbers "will never have a normal representation in the brain."

One of the recommendations of the neuroscientists conducting these important studies is that schools focus on *finger discrimination*—not only on number counting via their fingers but also on helping students distinguish between those fingers. Still, schools typically pay little if any attention to finger discrimination, and to our knowledge, no published curriculum encourages this kind of mathematical work. Instead, thanks largely to school districts and the media, many teachers have been led to believe that finger use is useless and something to be abandoned as quickly as possible. Kumon, for example, an after-school tutoring program used by thousands of families in dozens of countries, tells parents that finger-counting is a "no no" and that those who see their children doing so should report them to the instructor.

Stopping students from using their fingers when they count could, according to the new brain research, be akin to halting their mathematical development. Fingers are probably one of our most useful visual aids, and the finger area of our brain is used well into adulthood. The need for and importance of finger perception could even be the reason that pianists, and other musicians, often display higher mathematical understanding than people who don't learn a musical instrument.

Teachers should celebrate and encourage finger use among younger learners and enable learners of any age to strengthen this brain capacity through finger counting and use. They can do so by engaging students in a range of classroom and home activities, such as those shown in Figures 1 and 2. (The full set of activities can be found at this website: https://www.theatlantic.com/education/archive/2016/04/why-kids-should-use-their-fingers-in-math-class/478053/.

The finger research is part of a larger group of studies on cognition and the brain showing the importance of visual engagement with math. Our brains are made up of "distributed networks," and when we

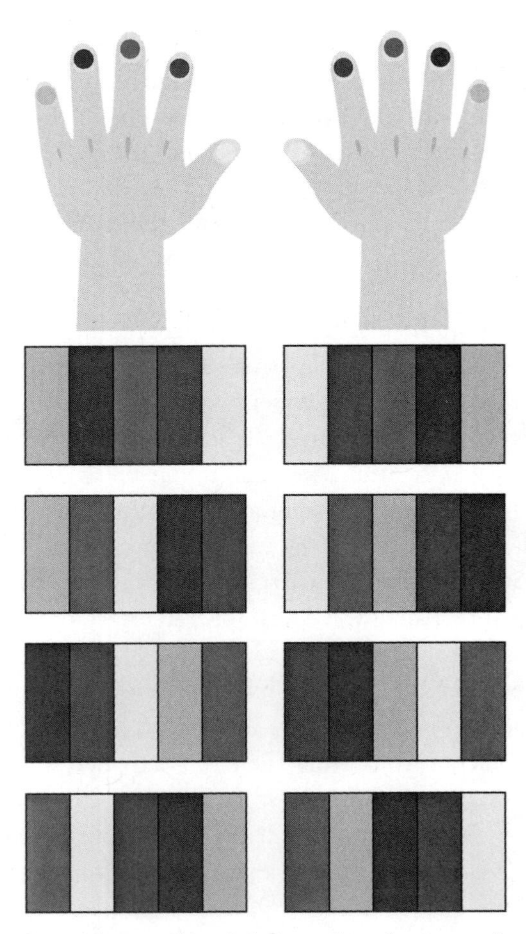

FIGURE 1. Give the students colored dots on their fingers and ask them to touch the corresponding piano keys.

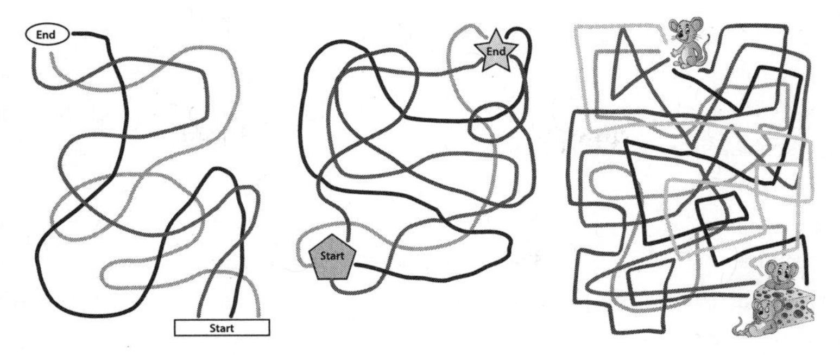

FIGURE 2. Give the students colored dots on their fingers and ask them to follow the lines on increasingly difficult mazes.

handle knowledge, different areas of the brain communicate with each other. When we work on math, in particular, brain activity is distributed among many different networks, which include areas within the ventral and dorsal pathways, both of which are visual. Neuroimaging has shown that even when people work on a number calculation, such as 12 × 25, with symbolic digits (12 and 25) our mathematical thinking is grounded in visual processing.

A striking example of the importance of visual mathematics comes from a study showing that after four 15-minute sessions of playing a game with a number line, differences in knowledge between students from low-income backgrounds and those from middle-income backgrounds were eliminated (Figure 3).

Number line representation of number quantity has been shown to be particularly important for the development of numerical knowledge, and students' learning of number lines is believed to be a precursor of children's academic success.

Visual math is powerful for all learners. A few years ago, Howard Gardner proposed a theory of multiple intelligences, suggesting that people have different approaches to learning, such as those who are visual, kinesthetic, or logical. This idea helpfully expanded people's thinking about intelligence and competence, but was often used in unfortunate ways in schools, leading to the labeling of students as particular types of learners who were then taught in different ways. But people who are not strong visual thinkers probably need visual thinking more than anyone. Everyone uses visual pathways when we work on math.

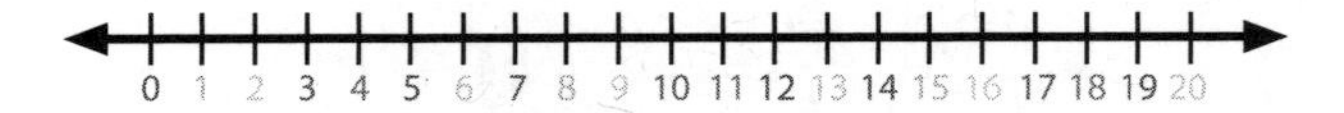

FIGURE 3. Give the students a colored number line and help them build visuo-spatial knowledge of number and quantity.

The problem is that it has been presented, for decades, as a subject of numbers and symbols, ignoring the potential of visual math for transforming students' math experiences and developing important brain pathways.

Related Story

It is hardly surprising that students so often feel that math is inaccessible and uninteresting when they are plunged into a world of abstraction and numbers in classrooms. Students are made to memorize math facts and plow through worksheets of numbers, with few visual or creative representations of math, often because of policy directives and faulty curriculum guides. The Common Core standards for kindergarten through eighth grade pay more attention to visual work than many previous sets of learning benchmarks, but their high school content commits teachers to numerical and abstract thinking. And where the Common Core does encourage visual work, it's usually encouraged as a prelude to the development of abstract ideas rather than a tool for seeing and extending mathematical ideas and strengthening important brain networks.

To engage students in productive visual thinking, they should be asked, at regular intervals, how they *see* mathematical ideas, and to draw what they see. They can be given activities with visual questions, and they can be asked to provide visual solutions to questions. When the youcubed team (a center at Stanford) created a free set of visual and open mathematics lessons for grades 3 through 9 last summer, which invited students to appreciate the beauty in mathematics, they were downloaded 250,000 times by teachers and used in every state across the United States. Ninety-eight percent of teachers said that they would like more of the activities, and 89% of students reported that the visual activities enhanced their learning of mathematics. Meanwhile, 94% of students said that they had learned to "keep going even when work is hard and I make mistakes." Such activities not only offer deep engagement, new understandings, and visual brain activity, but they also show students that mathematics can be an open and beautiful subject, rather than a fixed, closed, and impenetrable subject.

Some scholars note that it will be those who have developed visual thinking who will be "at the top of the class" in the world's new high-tech workplace, which increasingly draws upon visualization technologies and techniques, in business, technology, art, and science. Work on mathematics draws from different areas of the brain, and students need to be strong with visuals, numbers, symbols, and words—but schools are not encouraging this broad development in mathematics now. This is not because of a lack of research knowledge on the best ways to teach and learn mathematics; it is because that knowledge has not been communicated in accessible forms to teachers. Research on the brain is often among the most impenetrable for a lay audience, but the knowledge that is being produced by neuroscientists, if communicated well, may be the spark that finally ignites productive change in mathematics classrooms and homes across the country.

Threshold Concepts and Undergraduate Mathematics Teaching

Sinéad Breen and Ann O'Shea

1. Introduction to Threshold Concepts

The idea of a threshold concept emerged from a U.K. national research project (*Enhancing Teaching-Learning Environments in Undergraduate Courses*, 2001–2005) designed to support departments involved in undergraduate teaching in thinking about new ways of encouraging high-quality learning [5]. In pursuing this research in the field of economics, it became clear to Erik Meyer and Ray Land that certain concepts were held by economists to be essential to the mastery of their subject. These concepts were seen to have certain features in common and were called "threshold concepts" [16]. The notion of a threshold concept was introduced as a way of differentiating between learning outcomes that involved "seeing things in a new way" and those that did not. Threshold concepts have been described as portals, opening up a new and previously inaccessible view of a topic, a view without which students would be unable to fully progress intellectually. From that point of view, threshold concepts form a subset of what university lecturers would usually call "core concepts." A core concept is a conceptual building block: it must be understood but does not necessarily lead to a qualitatively different view of the subject matter.

Meyer and Land [16] originally identified five characteristics of a threshold concept: transformative, irreversible, integrative, bounded, and troublesome. What do they mean by these terms? Let us start with the idea that a threshold concept is *transformative*. Meyer and Land claim that once a threshold concept is understood, it has the potential to trigger a significant shift or transformation in the perception of a subject, or part thereof. The mastery of each threshold concept could be viewed as a step toward acquiring a professional's appreciation of the subject; this represents an ontological shift (or change in *being*) as well as a conceptual shift. The change in perception is unlikely to be forgotten and can be "unlearned" only with considerable effort; therefore, this concept is considered *irreversible*. For this reason, it can be difficult for lecturers or experienced practitioners to appreciate the difficulties of their students, as this requires them to look back over thresholds they have long since crossed. Threshold concepts often expose the interrelatedness of a topic and allow connections that were previously hidden to be displayed. They can bring different aspects of a subject together and act as an anchor for the subject. From this perspective, they have been described as *integrative*. Often, but not necessarily, threshold concepts may lie on the border between conceptual spaces or may constitute the demarcation line between disciplinary areas. For this reason, they have been described as *bounded*. Finally, threshold concepts are *troublesome*, in part because of the characteristics described above, but also because they are often inherently conceptually difficult. The concept may appear to be counterintuitive, paradoxical, or incoherent, or it may involve subtle distinctions being made between ideas.

In fact, Davies and Mangan [6] have argued that the transformative, integrative, and irreversible characteristics of a threshold concept are necessarily interwoven:

> A concept that integrates prior understanding is necessarily transformative, because it changes a learner's perception of their existing understanding. If a concept integrates a spectrum of prior understanding, it is more likely to be irreversible, because it holds together a learner's thinking about many different phenomena. To abandon such a threshold concept would be massively disruptive to an individual's whole way of thinking (p. 712).

2. *Threshold Concepts in Mathematics*

Let us consider some candidates for the title "threshold concept" from the undergraduate mathematics curriculum.

2.1. Limits

When considering the attributes of a threshold concept, most readers will likely be reminded of the problems that students encounter with the ϵ-δ definition of the limit of a function. Indeed, Meyer and Land included this as an example in their original work [16]. They remarked that

> In pure mathematics the concept of a limit is a threshold concept; it is the gateway to mathematical analysis and constitutes a fundamental basis for understanding some of the foundations and application of other branches of mathematics such as differential and integral calculus (p. 3).

Understanding the limit definition opens the door to the field of analysis and sits on the boundary between calculus and analysis courses. (It could thus be thought of as bounded, using the terminology of [16].)

The concept is certainly a troublesome one for most students, and this is not surprising since, historically, the evolution of the notion was slow. Even though Newton and Leibniz developed calculus in the 17th century, and some of the ideas had previously been in use for a long time, it was not until the 19th century that Weierstrass finally formulated the ϵ-δ definition [1, p. 287].

There has been a considerable amount of research into the problems that students face with the notion of a limit. These problems could be divided into two main categories: those that arise from preexisting images of limits and those that stem from the formulation of the definition itself [19]. Research has shown ([4], [18]) that the images that students have that relate to the word "limit" can affect and inhibit their understanding of the concept when they meet it in an analysis course. Cornu [4] remarked that in the case of limits, both the phrase "tends to" and the word "limit" have interpretations in everyday life that are not always consistent with their mathematical meanings. For example, it carries the connotation of an impassable limit that is impossible to reach, a maximum or minimum, or a finishing point; each of these conceptions can cause problems for students, even after they are introduced to the rigorous definition. On the other hand, the structure of the definition itself causes problems. First, it contains the quantifiers $\forall$ and $\exists$, which together prove confusing to students. Also, students often fail to see how the existence of a limit $\lim_{x \to a} f(x) = L$ can be inferred from a statement about inequalities such as $\forall \epsilon > 0$, $\exists \delta > 0$ such that $|f(x) - L| < \epsilon$ if $0 < |x - a| < \delta$ [19]. Students seem to want a formula or algorithm with which they can compute the limit ([18], [19]), and they are uncomfortable with using the definition instead.

Anecdotally, students and mathematicians often report on the moment when the point of the ϵ-δ definition became clear to them. The fact that they can remember a precise moment when this happened is significant and points to the transformative and irreversible nature of the new understanding.

Research has found ([23]) that further difficulties in understanding limits may arise from a mismatch between the (formal) concept definition and students' concept image. Tall and Vinner ([23] p. 151) defined the notion of a concept image as consisting of "all the cognitive structure in the individual's mind that is associated with a given concept." They found that for the topic of limits of functions, students' concept images may contain elements that do not agree with the definition or even with other parts of the concept image. Przenioslo [18] studied the conceptions of limit held by undergraduate students. She found that students had images of limits that were based on the formal definition, on the computation of limits using algorithms, on the dynamic nature of limits (i.e., thinking of values approaching a certain point), and on the function value at a point. She conjectured that the last three images were based on informal definitions used previously. In her study, the students whose images were close to the definition were usually more successful than the others at solving problems about limits, but also rarely reverted back to language such as "getting closer and closer." This also provides evidence that understanding the definition is a transformative and irreversible experience. This may be one reason why it is difficult to teach this topic, since once one has crossed the threshold with the limit definition, it is difficult to remember what it was like on the other side.

2.2. Functions

Even before students encounter the ϵ-δ definition of the limit of a function, of course, they will have worked with the concept of a function. This concept is fundamental in modern mathematics, and even though students are exposed to this idea in school, it has been found that many undergraduates have difficulties with it [3]. Pettersson [17] has suggested that the concept of function is a threshold concept.

The mathematics education community has conducted many studies into students' understandings of, and difficulties with, functions. Once again, as was the case with limits, we find that one of the main problems that students face is that of the definition. For example, Vinner and Dreyfus [25] found that students often think of functions as being a formula or an equation, and may be loath to accept functions that are not defined by a single algebraic expression. They may also expect all functions to be continuous. These problems with the definition of a function bear similarities to the stages of the historic development of the concept [14], and so make a case for the concept to be described as inherently conceptually difficult (or troublesome).

Perhaps as a consequence of viewing functions as defined by an algebraic expression, students often think of them in terms of actions or an input–output model. For example, they may see $f(x) = 5x - 2$ as a recipe for a series of calculations, rather than as an object in its own right. To properly understand functions, and to work with them in diverse areas of mathematics, students should be able to conceive of a function as an action, as a process, and as an object [2]. Sfard [22] discusses the complementary approaches of dealing with abstract notions such as functions: operationally as processes and structurally as objects. She introduced the term "reification" to represent the transition of thought involved when a learner progresses to viewing processes as objects. She warns that reification is an "ontological shift, a sudden ability to see something familiar in a new light" (p. 19) and a "rather complex phenomenon" (p. 30), causing obstacles and frustration for learners; this reinforces a view of the concept as troublesome and illustrates how reification can be viewed as transformative. Gray and Tall [10] maintain that the ability to think flexibly in this manner (operationally and structurally) is at the root of successful mathematical thinking. They also suggest that the flexibililty in thought achieved by those who have experienced reification can explain why a mathematics expert may find it difficult to appreciate the difficulties of a novice, pointing to an irreversibility, as described in [16]. Reification seems to be quite similar to what Thurston [24] called "compression." He spoke about learners of mathematics working step by step and struggling to understand a concept, but asserted that, once they have really understood the concept, their perspective can change to being able to see it as a whole. He believed that such insight and mental compression can make it easier to recall and use the idea when it is needed in future.

Finally, the idea of a function permeates many areas of mathematics, and as such, a comprehensive understanding of the concept can expose previously hidden connections between different topics. Students usually first meet the formal definition of a function in the context of analysis, but once it is properly understood, they often come to realize how it can be related to linear systems and matrices they have encountered in algebra, for instance. In this sense, it could be described as integrative.

2.3. Cosets and Quotient Groups

In contrast with the concepts of functions and limits, relatively little research has been carried out into the teaching and learning of abstract algebra. However, researchers have suggested that students' difficulties in abstract algebra courses seem to deepen when they meet the concepts of cosets and quotient groups ([9], [13]). These concepts are crucial to the study of group theory, and so they could present an obstacle to further progression in algebra. In a study of second-year undergraduate students at a British university, Ioannou [13] reported that students had problems visualizing cosets. This issue led to students encountering problems understanding the remainder of their group theory course and also contributed to diminishing levels of engagement with the course.

Dubinsky et al. [9] found that students were more comfortable with cosets when they could form them by carrying out calculations—that is, by performing an action or following a process. However, they had difficulties when faced with the formation of cosets in unfamiliar settings. There was evidence that some students in that study saw cosets only in terms of an action or process to be carried out, rather than as objects in their own right. This limitation led to difficulties when thinking about cosets as elements of a quotient group. Students who could view cosets as objects were better able to answer difficult questions on the topic, and so, once this reification took place, it seemed to be transformative and probably irreversible.

3. Threshold Concepts and Implications for Undergraduate Mathematics Teaching

How can such research in mathematics education and the identification of threshold concepts inform the practice of mathematics teaching and learning? In "An introduction to threshold concepts," Cousin [5] claims that a

> tendency among academic teachers is to stuff their curriculum with content, burdening themselves with the task of transmitting vast amounts of knowledge bulk and their students of absorbing and reproducing this bulk (p. 4).

Criticism has been leveled at mathematics lecturers, in particular, for such a practice: Hillel [12] claims that, generally speaking, undergraduate mathematics courses have been defined in terms of mathematical content and the techniques students are expected to master or theorems they should be able to prove. Although the main goal of a mathematics lecturer may be to foster mathematical understanding in their students, such an understanding is seldom specifically nurtured by the mathematical tasks and assessments students are required to complete [20], leading many authors to decry an overemphasis on procedures and the reproduction of definitions, statements of theorems, etc., in undergraduate mathematics modules. Consequently, this overemphasis can result in a reliance on shallow, superficial, or rote learning by students and an inability to answer unseen problems or to apply or transfer their mathematical knowledge as appropriate [21]. For instance, Dreyfus [8] asserts that many students learn a large number of standardized procedures in their university mathematics courses, and, although they end up with a considerable amount of mathematical knowledge, they cannot use it in a flexible manner:

> They have been taught the products of the activity of scores of mathematicians in their final form but they have not gained insight into the processes that have led mathematicians to create these products (p. 28).

This is very much in contrast with the type of approach advocated by Land et al. [15], who suggest that a focus on threshold concepts can enable teachers to make refined decisions about what is fundamental to the study and mastery of their subject. Because of the potentially powerful transformative effects of threshold concepts on the learning experience, they advocate treating threshold concepts as "jewels in the curriculum," around which courses could be organized. In addition, since a poor understanding of these concepts can form a barrier to further advancement, they should be given particular attention when designing the curriculum. If we, as mathematicians, can identify these concepts, we may be able to help give students both the tools and the time they need to develop a mastery of them. This curriculum design may involve a recursive (as opposed to a linear) approach, revisiting threshold concepts at various stages and from various perspectives throughout a module or program. Land et al. [15] advocate that a framework of engagement should be constructed by lecturers to facilitate the development of students' understanding of threshold concepts, actively engaging students with the conceptual material and allowing students to experience the "ways of thinking and practicing" that are expected of practitioners in their discipline. In particular, they recommend that

> tutors ask students to explain [a troublesome concept], to represent it in new ways, to apply it to new situations, to connect it to their lives. The emphasis is equally strong that they should not simply recall the concept in the form in which it was presented (p. 57).

Teachers should be cautious when making assumptions about what students' uncertainties might be. As mentioned earlier, it can be difficult for experienced teachers to understand the obstacles met by students as they grapple with a difficult concept for the first time. Indeed, Thurston [24] (although he was not speaking about threshold concepts) also made this point and remarked that once one has mastered concepts it is very hard to "put oneself back in the mind of someone to whom they are mysterious" (p. 848). This change in perception "puts a psychological barrier in the way of listening fully to students" ([24], p. 848). Land et al. [15] advise lecturers to listen not just for what students know, but also for the terms that shape their knowledge and define their uncertainties and instabilities.

Land et al. [15] also discuss the "indispensable role of metacognition in the learning process" (p. 59). They outline how lecturers should empathize with learners who are grappling with troublesome concepts, make sure that they are aware that others are experiencing similar difficulties, and encourage them to tolerate uncertainty in the short term. Students often abandon their studies because of conceptual difficulties, not realizing that the confusion they are experiencing may be short-lived. It has also been suggested that students may be more likely to resort to mimicry or plagiarism if they feel that they are alone in their confusion [5]. Furthermore, making students aware of the historical development of concepts may be useful not only in encouraging engagement with a concept, but also in allowing them to appreciate the difficulties experienced by those responsible for first articulating or formulating a concept, thereby encouraging perseverance.

We have seen that, for many threshold concepts, reification is an important part of the development of understanding and thus can serve as a marker of students' progress in learning mathematics. In mathematics teaching, however, reification often remains an implicit learning outcome, a form of tacit knowledge that is not explicitly articulated to learners. It may be that by focusing on threshold concepts in the curriculum, this process of reification can be addressed in a more explicit manner.

Some studies have been undertaken attempting to put these recommendations into practice. Harlow et al. [11] outline findings from a collaborative action-research project to document changes in lecturers' threshold-concept-informed teaching and their effect on student learning in analog electronics. The lessons for teachers learned through this project are described as listening to students, tolerating learner confusion, and revisiting threshold concepts, echoing the recommendations given by Land et al. [15]. Davies and Mangan [7] have also endeavored to put theory into practice in constructing a "framework of engagement" for first-year undergraduate economics students. They blended insights from the theory of threshold concepts and variation theory to propose four pedagogic principles, which were then translated into three types of teaching and learning activity—reflective exercises, problem-focused exercises, and threshold network exercises—and they report on their experiences of using these activities.

From a mathematics perspective, although Dubinsky et al. [9] do not frame their discussions of teaching group theory in general, and cosets in particular, in terms of threshold concepts, they make pedagogical recommendations in line with those described above and report some success from their efforts. For instance, they suggest "finding alternatives to linear sequencing" of material and state that "it is the role of the teacher, not to eliminate [students'] frustration, but to help students learn to manage it" (p. 300). In particular, using technology, they aim to help students experience reification by moving from viewing cosets in terms of actions to seeing them as objects.

4. Concluding Remarks

In this article, we have described what is meant by a threshold concept, given examples of some mathematical concepts that have been identified as threshold concepts, and discussed how they could be used in teaching and especially in curriculum design. In summary, lecturers should give special attention to threshold concepts and use them as a central motif for courses; they should revisit the concepts frequently and view them from different perspectives if possible; they should become familiar with the literature on student misconceptions in order to help understand what difficulties students might face; they could make students aware that having difficulty understanding these concepts is common but not insurmountable. We have found the idea of a threshold concept and these recommendations both interesting and useful in developing our own teaching practice; we hope the wider mathematical community will do likewise. A comprehensive survey of research undertaken on threshold concepts can be found at http://www.ee.ucl.ac.uk/mflanaga/thresholds.html.

References

1. Bowyer, C. B. 1949. *The History of the Calculus and Its Conceptual Development*. New York, Dover Publications.
2. Breidenbach, D., E. Dubinsky, J. Hawks, and D. Nichols. 1992. "Development of the process conception of function." *Educational Studies in Mathematics*. 23(3): 247–285.
3. Carlson, M. 1998. "A cross-sectional investigation of the development of the function concept." In A. H. Schoenfeld, J. Kaput, and E. Dubinsky (Eds.), *Research in Collegiate Mathematics Education III*. CBMS Issues in Mathematics Education (pp. 114–162). Providence, RI, American Mathematical Society.
4. Cornu, B. 1992. "Limits." In D. Tall, (Ed.), *Advanced Mathematical Thinking* (pp. 153–166). Dordrecht, Netherlands, Kluwer Academic Publishers.
5. Cousin, G. 2006. "An introduction to threshold concepts." *Planet*. 17: 4–5.

6. Davies, P., and J. Mangan. 2007. "Threshold concepts and the integration of understanding in economics." *Studies in Higher Education*. 32(6): 711–726.
7. Davies, P., and J. Mangan. 2008. "Embedding threshold concepts: From theory to pedagogical principles to learning activities." In R. Land, J. H. F. Meyer, and J. Smith (Eds.), *Threshold Concepts in the Disciplines* (pp. 37–50). Rotterdam, Netherlands, Sense Press.
8. Dreyfus, T. 1991. "Advanced mathematical thinking processes." In D. Tall (Ed.), *Advanced Mathematical Thinking* (pp. 25–41). Dordrecht, Netherlands, Kluwer Academic Publishers.
9. Dubinsky, E., J. Dautermann, U. Leron, and R. Zazkis. 1994. "On learning fundamental concepts of group theory." *Educational Studies in Mathematics*. 27(3): 267–305.
10. Gray, E., and D. Tall. 1994. "Duality, ambiguity and flexibility: A 'proceptual' view of simple arithmetic." *Journal for Research in Mathematics Education*. 25(2): 116–140.
11. Harlow, A., M. Peter, J. Scott, and B. Cowie. 2012. *Students' perceptions of travel through liminal space: Lessons for teaching*. http://www.nairtl.ie/documents/EPub_2012Proceedings.pdf. Accessed May 5, 2015.
12. Hillel, J. 2001. "Trends in curriculum: A working group report." In D. Holton (Ed.), *The Teaching and Learning of Mathematics at University Level* (pp. 59–70). Dordrecht, Netherlands, Kluwer Academic Publishers.
13. Ioannou, M. 2010. "Visualisation of cosets and its impact on student engagement with group theory." In M. Joubert and P. Andrews (Eds.), *Proceedings of the British Congress for Mathematics Education* (pp. 259–264). Manchester, U.K., BSRLM.
14. Kleiner, I. 2012. *Excursions in the History of Mathematics*. Basel, Switzerland, Birkhauser.
15. Land, R., G. Cousin, J. H. F. Meyer, and P. Davies. 2005. "Threshold concepts and troublesome knowledge (3): Implications for course design and evaluation." In C. Rust (Ed.), *Improving Student Learning Diversity and Inclusivity* (pp. 53–64). Oxford, U.K., OCSLD.
16. Meyer, J. H. F., and R. Land. 2003. "Threshold concepts and troublesome knowledge 1—Linkages to ways of thinking and practising." In C. Rust (Ed.), *Improving Student Learning—Ten Years On* (pp. 412–424). Oxford, U.K., OCSLD.
17. Pettersson, K. 2010. *Threshold concepts: A framework for research in mathematics education*. http://www.cerme7.univ.rzeszow.pl/index.php?id=wg14. Accessed May 5, 2015.
18. Przenioslo, M. 2004. "Images of the limit of function formed in the course of mathematical studies at the university." *Educational Studies in Mathematics*. 55: 103–132.
19. Roh, K. H. 2010. "An empirical study of students' understanding of a logical structure in the definition of limit via the ϵ-strip activity." *Educational Studies in Mathematics*. 73: 263–279.
20. Sangwin, C. 2003. "New opportunities for encouraging higher level mathematical learning by creative use of emerging computer assessment." *International Journal of Mathematical Education in Science and Technology*. 34(6): 813–829.
21. Schoenfeld, A. H. 1989. "Explorations of students' mathematical beliefs and behavior." *Journal for Research in Mathematics Education*. 20: 338–355.
22. Sfard, A. 1991. "On the dual nature of mathematical conceptions: Reflections on processes and objects as different sides of the same coin." *Educational Studies in Mathematics*. 22(1): 1–36.
23. Tall, D., and S. Vinner. 1981. "Concept image and concept definition in mathematics with particular reference to limits and continuity." *Educational Studies in Mathematics*. 12: 151–169.
24. Thurston, W. P. 1994. "Mathematical education." *Notices of the AMS*. 37(7): 844–850.
25. Vinner, S., and T. Dreyfus. 1989. "Images and definitions for the concept of function." *Journal for Research in Mathematics Education*. 20(4): 356–366.

Rising above a Cause-and-Effect Stance in Mathematics Education Research

JOHN MASON

René Descartes was strongly influenced by the cuckoo clock: the idea of a mechanism with logical and necessary consequences as a metaphor for human experience. Pierre-Simon Laplace applied this mathematically by claiming that if he had all the initial conditions at some point in time, then he could predict everything subsequently. The cuckoo clock mechanism metaphor, complete cause and effect, underpinned the Age of Enlightenment and the Industrial Revolution because mechanisms can be engineered to be independent of time, place, situation, and conditions.

So why is this aspect of Descartes and Laplace still influential in mathematics education? Why does so much research in mathematics education keep chasing a *will o' the wisp*, pursuing a magic ring whose presence would provide a primary explanation and predictor of effective teaching and proficient learning? This chase continues even though it is commonly acknowledged that learning is highly contingent and that mathematics learning is especially so. It is influenced by time, place, situation, conditions, dispositions, recent and long-term past experience, colleagues, the topic, pedagogical actions, ... The list continues.

Nevertheless, people continue to ask questions such as "What does effective teaching of mathematics look like?" and "What must teachers know in order to teach effectively?" They do this despite rich experience that what "knowledge" teachers have displayed in the past, what actions they have written essays about or even enacted in the past, does not enable prediction of what actions will be enacted in the future. Even shared development of a lesson does not lead to the same lesson being enacted, because the people and the situation are subtly different (Kuhlberg 2007).

It is strange that in a world in which attempts to construct walking robots by controlling all their components failed, whereas allowing feedback loops to operate enables the mechanism as a whole to walk reasonably naturally, we persist in trying to control things from the center. Industry, and certainly education, retains the pyramid structure of power and control. The 1980s slogan of accountability, which now so clogs our institutions with form-filling and back-watching, has been activated through central control: each person in the chain of command feels the need to be assured that those carrying out policy for them are behaving appropriately. Organic growth and transformation are stymied. The result is stagnation and fear of innovation. Yet at the same time, people cry out for innovative practices in education, problem-solving, creativity, and the like, as if somehow an innovative practice injected into a moribund system is likely to enliven it.

We desperately need research that sees and conceives of the classroom as an organic whole, in which both individuals and the collective display qualities of complexity (life). This research will necessarily include seeing and conceiving of the educational institution as itself an organism, within an educational nexus, within a state-guided system. Such a stance could not only bring freshness to how we see education, but could also release much desired creativity and liveliness. This is the stance of complexity theorists (Davis et al. 2006).

As a domain of study, mathematics education has become proud of its multidisciplinary approach, drawing on (the foci of, the fads of) other disciplines. These disciplines sweep over one another like waves on a beach. Constructivisms (of various sorts), constructionism, and theories with adjectives such as sociocultural, sociohistorical, political, critical, identity, positioning, attribution, situated cognition, and variation have all taken the limelight for a period of time, then have proved inadequate and so been abandoned. The waves keep breaking on the beach, attracting attention, then being overlaid, partly because they fail as universal panacea, and partly because it is easier to ride a fresh wave than to probe deeply into the consequences of an earlier wave.

Learning and teaching are fundamentally relational. My relation with my students and my relation with mathematics interweave to influence how I am with students and colleagues and so to afford what is possible (Mason 2008, 2014). These relationships coevolve, because responsiveness to student concerns and states influences what can be said and done, and what is said and done reciprocally influences the state of individuals and the state of the class as a whole. They are not only complex, but also full of complexity, in the full sense of the term: multiple factors and components in a complex web of interrelatedness. They involve development, coemergence, and variety.

Every teacher act, every student reaction, refeeling, response, every shift of attention is part of the transformation of the situation that involves teacher, student, and mathematics within the milieu of classroom, institution, and education system. What looks like the same pedagogical act enacted by a teacher in two different situations can lead to widely varying attributes of the system in both the short and the long term. This notion is characteristic of complexity. Terms used to capture that complexity (Davis et al. 2006) include fractal, recursive, and self-organizing or autopoetic (Maturana and Varela 1972, 1988).

Taking a systemic stance retains the complexity rather than trying to reduce phenomena to simple components whose qualities when analyzed independently may or may not be present in the full system. Instead of seeking illusive stable or robust outcomes from micromanaging a few components, a systemic approach recognizes that it may or may not be possible to identify “strange attractors” that are redolent of certain types of complex situations.

For example, documenting the deficiencies of novice teachers and students simply contributes to the general malaise of teaching and learning mathematics as inadequate and ineffective. Tracking changes in learner proficiency over time, or teacher use of pedagogic actions over time underpins a linear trail of development rather than elaborating the complexity of change and development, of what is possible in the moment or on the fly. Even the terms “growth” and “development” carry with them the frozen metaphor of “progress as a staircase” rather than contributing to a metaphor of “emergence of complexity,” of “folding back” (Pirie and Kieren 1989, 1994) so that

> We shall not cease from exploration
> And the end of all our exploring
> Will be to arrive where we started
> And know the place for the first time.
>
> —T. S. Eliot, *Little Gidding*

One of the contributions of the discipline of noticing (Mason 2002) is that it circumvents a cause-and-effect stance. It offers practical techniques for noticing opportunities in the moment to act freshly rather than habitually. It is not about accumulating a collection, or worse, a sequence of pedagogical acts that are claimed to have specific effects. Rather it is a systematic method of enriching the affordances and the repertoire of available actions so that in-the-moment choices are available. It is about recognizing and seizing opportunities to act freshly, rather than being condemned to act mechanically out of habit. It is not about finding the "best action" to initiate when students act, emote, or respond in a particular way, but about enabling a choice to be made, a momentary taste of freedom of action, consonant with the teacher's apprehension of the situation in all its complexity. The more narrow and tunnel-visioned the awareness of the teacher, the fewer the actions that are available to be enacted, the simpler the situation as perceived and apprehended, and so the greater the chance that complexity is lost.

Repeatedly refining distinctions (for example, as is currently happening in studies of enculturating learners into mathematical reasoning) can only be of value if it contributes to a refinement of how we sense the complex web of interdependencies and relationships, rather than implying that a sufficiently refined set of distinctions will somehow make it possible to engineer suitable acts of teaching whose outcome will be enhanced proficiency for learners. As Davis (2014) points out, even the notion of "outcome" invokes the metaphor of input–output and mechanicality.

The human psyche is made up not only of enaction, affect, and cognition, but also of attention, will, and witness. Preparing a lesson or sequence of lessons is about bringing to mind the web of possible shifts of attention that participants might experience (not simply what is attended to, but also in what ways). Teachers' choices will not cause learners to experience something, but they can enrich possibilities, the opportunities for learners. "I cannot do the learning for my students." In other words, student will is significant: taking initiative is a necessary and unavoidable part of learning. Tasks can help prepare learners to be able to hear and see what is being said and done, with and in front of them. Different students may experience different trajectories through the teacher's perception of the web of interrelationships that constitute the experienced complexity, but with care, suitable constraints can be imposed so that attention does not simply go "everywhere." A successful lesson in mathematics is not one that could or does "go anywhere," but one in which the teacher has a rich collection of resources to invoke, depending on their sense of the topic and on what they are sensitized to notice happening, much of which can be anticipated.

I have long sought an alternative to cause and effect as a suitable mechanism for describing education. The best I have been able to find is the metaphor of "primordial ooze," or if you prefer, a complex chemical soup. There are multiple factors (chemicals) at play, with a dynamic consisting of combining and breaking up of identifiable or discernible components. Overall, there may be a dynamic equilibrium (a "strange attractor"), or there may be casual or abrupt drift of some sort.

One weakness of the metaphor is that the notion of chemicals suggests atomic components, whereas when probed, the analogue to chemicals are flows of energy along neural networks altering the attributes of those networks, and at a larger grain size, aspects of the human psyche being energized in broadly characteristic yet situationally specific ways. But further analysis of the human psyche reveals further complexity, in that, for example, cognition, affect, and enaction are themselves complex interrelated formations (Ouspensky 1950). It is only recently that the human psyche has been acknowledged to go beyond cognition, affect, and enaction to include attention, will, and witness. The latter list can be developed through practices such as those in the discipline of noticing.

I look to mathematics education to catch up with the zeitgeist of the times, to confine to the filing cabinet studies of what people can and cannot do, even when the studies are more appropriately cast as "did and did not do." Let us focus attention on the web of influences on specific classroom moments, being content for the moment to chart their multiplicity, and to reach out for techniques for studying complex systems. Charting the presence of feedback loops that manifest structural coupling (Maturana and Varela 1988) within the web could enrich teacher awareness, providing teachers with yet further actions to initiate in response to learners' actions. Detailed studies of components make up one contribution, but they cannot be effectively acted upon, cannot be effectively used to influence policy unless they are seen as part of a much greater whole.

References

Davis, B. (2014). *EDUC 420 Lectures*. Calgary: Werklund School of Education, University of Calgary.

Davis, B., Summara, D., and Simmt, E. (2006). *Complexity and education: Inquiries into learning, teaching and research*. Mahwah, NJ: Lawrence Erlbaum.

Kuhlberg, A. (2007). "Can lessons be repeated?" In J. H. Woo, H. C. Lew, K. S. Park, and D. Y. Seo (Eds.), *Proceedings of the 31st conference of the international group for the psychology of mathematics education* (Vol. 3, pp. 121–128). Seoul: PME.

Mason, J. (2002). *Researching Your Own Practice: The Discipline of Noticing*. London: RoutledgeFalmer.

Mason, J. (2008). "Being mathematical with & in front of learners: Attention, awareness, and attitude as sources of differences between teacher educators, teachers & learners." In T. Wood (Series Ed.) & B. Jaworski (Vol. Ed.), *International Handbook of Mathematics Teacher Education: The Mathematics Teacher Educator as a Developing Professional* (Vol. 4, pp. 31–56). Rotterdam, Netherlands: Sense Publishers.

Mason, J. (2014). *On being mathematical with and in front of learners*. Calgary: Killam Scholar Lecture, Werklund School of Education, University of Calgary.

Maturana, H., and Varela, F. (1972). *Autopoesis and cognition: The realization of the living*. Dordrecht, Netherlands: Reidel.

Maturana, H., and Varela, F. (1988). *The tree of knowledge: The biological roots of human understanding*. Boston: Shambala.

Ouspensky, P. (1950). *In search of the miraculous: Fragments of an unknown teaching*. London: Routledge & Kegan Paul.

Pirie, S., and Kieren, T. (1989). "A recursive theory of mathematical understanding." *For the Learning of Mathematics, 9*(4), 7–11.

Pirie, S., and Kieren, T. (1994). "Growth in mathematical understanding: How can we characterise it and how can we represent it?" *Educational Studies in Mathematics, 26*(2–3), 165–190.

How to Find the Logarithm of Any Number Using Nothing But a Piece of String

Viktor Blåsjö

The shape of a freely hanging chain suspended from two points is called the catenary, from the Latin word for chain. In principle, any piece of string would do, but one speaks of a chain since a chain with fine links embodies in beautifully concrete form the ideal physical assumptions that the string is nonstretchable and that its elements have complete flexibility independent of each other.

In modern terms, the catenary can be expressed by the equation $y = (e^x + e^{-x})/2$. As we shall see, Leibniz did not state this formula explicitly, but he understood well the relation it expresses, calling it a "wonderful and elegant harmony of the curve of the chain with logarithms" [6, p. 436]. (English translations of [5] and [6] are given in [10].) Indeed, he continued, the close link between the catenary and the exponential function means that logarithms can be determined by simple measurements on an actual catenary. "This may be helpful since during long journeys one may lose one's table of logarithms. . . . In case of need the catenary can then serve in its place" [7, p. 152]. Leibniz's recipe for determining logarithms in this way is delightfully simple and can easily be carried out in practice using, for example, a cheap necklace pinned to a cardboard box with sewing needles.

Leibniz's Recipe

Refer to Figure 1 and the following description:

(a) Suspend a chain from two horizontally aligned nails. Draw the horizontal through the endpoints, and the vertical axis through the lowest point.
(b) Put a third nail through the lowest point and extend one half of the catenary horizontally.
(c) Connect the endpoint to the midpoint of the drawn horizontal, and bisect the line segment. Drop the perpendicular through this point, draw the horizontal axis through the point where the perpendicular intersects the vertical axis, and take the distance from the origin of the coordinate system to the lowest point of the catenary to be the unit length. We will show below that the catenary now has the equation $y = (e^x + e^{-x})/2$ in this coordinate system.
(d) To find $\log(Y)$, find $(Y + 1/Y)/2$ on the y-axis and measure the corresponding x-value (on the catenary returned to its original form). This assumes that $Y > 1$. To find logarithms of negative values, use the fact that $\log(1/Y) = -\log(Y)$. If you seek the logarithm of a very large value, then you may end up too high on the y-axis; in such cases, you can either try hanging the endpoints closer together or using logarithm laws to express the desired logarithm in terms of those of lower values.

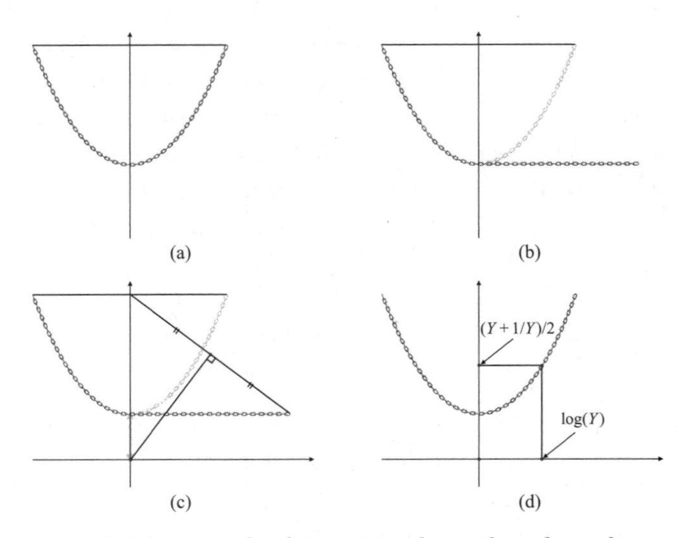

FIGURE 1. Leibniz's recipe for determining logarithms from the catenary.

The last step in this construction is given in Leibniz's catenary papers [5, 7, 8], where, however, the preceding steps are implicit at best; Leibniz later spelled these steps out in [11, No. 199].

The validity of this construction may be confirmed as follows. Figure 2 shows the forces acting on a segment of a catenary starting from its lowest point: the tension forces at the endpoints, which act tangentially, and the gravitational force, which is proportional to the arc s from T_0 to T. Since the catenary is in equilibrium, it is evident that the horizontal and vertical components of T balance with T_0 and the weight as, respectively, so $T_x = -T_0$ and $T_y = -as$. But since T acts in the direction of the tangent, we also know that $T_y/T_x = dy/dx$. Thus we obtain $dy/dx = as/T_0$. On the left half of the catenary, where s is negative, we get instead $T_y = as$ and $-T_y/T_x = dy/dx$, which gives the same result. For convenience, we choose the units of force and mass so that $a/T_0 = 1$, which gives $dy/dx = s$ as the differential equation for the catenary. Squaring both sides of this equation and using the Pythagorean identity $(dx)^2 + (dy)^2 = (ds)^2$ to eliminate dx leads to $(dy)^2 = s^2(ds^2 - dy^2)$ or $(1 + s^2)(dy)^2 = s^2(ds)^2$ and, by separating the variables and taking square roots,

$$dy = \frac{s\,ds}{\sqrt{1+s^2}},$$

which integrates to $y = \sqrt{1+s^2}$. Thus $s = \sqrt{y^2 - 1}$, which we can substitute into the original differential equation for the catenary to obtain $dy/dx = \sqrt{y^2 - 1}$. It is now straightforward to check that $y = (e^x + e^{-x})/2$ is the solution to this differential equation that passes through (0, 1).

It remains to verify that the coordinate system assumed in this solution is the same as that defined by the construction of Figure 1.

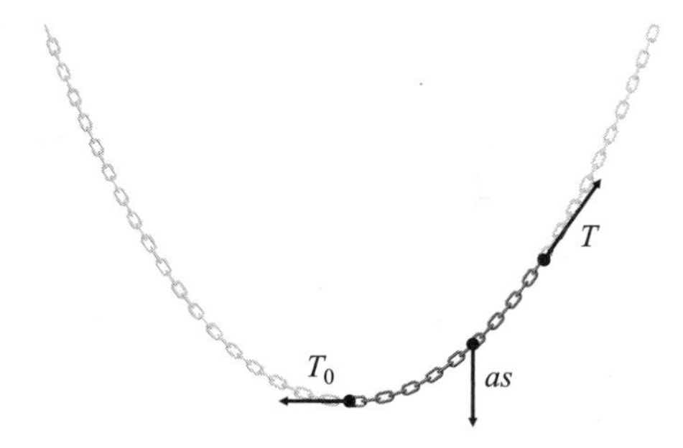

FIGURE 2. The forces acting on a segment of a catenary.

The key to Leibniz's verification turns out to be the intermediate step $y = \sqrt{1+s^2}$ above. To see this, consider Figure 3, which is Figure 1(c) with additional notation. We know from above that the catenary *FAL* is given by $y = (e^x + e^{-x})/2$ in a certain coordinate system whose origin O is at a vertical distance $OA = 1$ below the lowest point of the catenary. Consider the particular y-value $OH = y$ and the associated arc $AL = s$, then construct the horizontal segment AM with the same length s. It follows by the Pythagorean theorem that $OM = \sqrt{1+s^2}$. But above we saw that $y = \sqrt{1+s^2}$, which means in terms of this figure that $OH = OM$. Thus, OHM is an isosceles triangle, and so the perpendicular bisector of its base HM passes through the vertex O. This shows that the construction of Figure 1 does indeed give a way of recovering the coordinate system associated with the solution $y = (e^x + e^{-x})/2$, as we needed to show. From here it is a simple matter of algebra to check the final step of Figure 1.

In a Seventeenth-Century Context

Finding logarithms from a catenary may seem like an oddball application of mathematics today, but to Leibniz it was a very serious matter—not because he thought this method so useful in practice, but because it pertained to the very question of what it means to solve a

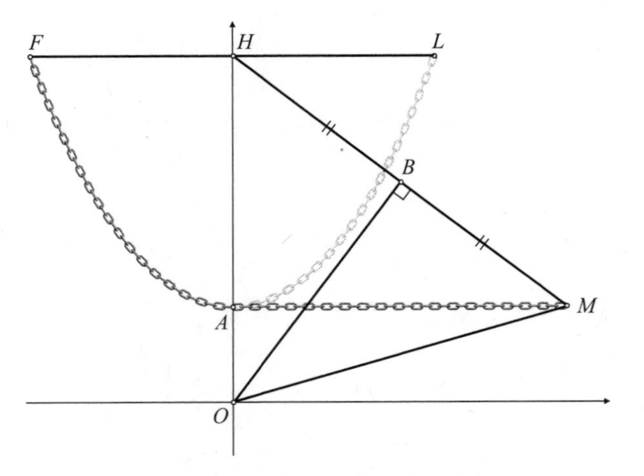

FIGURE 3. Figure used by Leibniz [11, No. 199] to justify the construction shown in Figure 1.

mathematical problem. Today we are used to thinking of a formula such as $y = (e^x + e^{-x})/2$ as the answer to the question of the shape of the catenary, but this would have been considered a naive view in the seventeenth century. Leibniz and his contemporaries discovered this relation between the catenary and the exponential function in the 1690s, but they never wrote this equation in any form, even though they understood perfectly well the relation it expresses. Nor was this for lack of familiarity with exponential expressions, at least in Leibniz's case, as he had earlier used such expressions to describe curves with considerable facility [11, No. 6].

Why, indeed, should one express the solution as a formula? What kind of solution to the catenary problem is $y = (e^x + e^{-x})/2$, anyway? The seventeenth-century philosopher Thomas Hobbes once quipped that the pages of the increasingly algebraical mathematics of the day looked "as if a hen had been scraping there" [4, p. 330], and what indeed is an expression such as $y = (e^x + e^{-x})/2$ but some chicken scratches on a piece of paper? It accomplishes nothing unless e^x is known already, i.e., unless e^x is more basic than the catenary itself. But is it? The fact that it is a simple formula of course proves nothing; we could just as well make up a symbolic notation for the catenary and then express the exponential function in terms of it. And, however one thinks of the graph of e^x, it can hardly be easier to draw than hanging a chain from two nails. So why not reverse the matter and let the catenary be the basic function and e^x the application? Modern tastes may have it that pure mathematics is primary and its applications to physics secondary, but what is the justification for this hierarchy? Certainly none that would be very convincing to a seventeenth-century mind.

The seventeenth-century point of view also had the authority of tradition on its side. Euclid's *Elements* had been the embodiment of the mathematical method for two millennia, and one of its most conspicuous aspects is its insistence on constructions. Euclid never proves anything about a geometrical configuration that he has not first shown how to construct by ruler and compass. These constructions are what gave meaning to mathematics and defined its ontology. This paradigm remained as strong as ever in the seventeenth century. When Descartes introduced analytic geometry in his *Géométrie* of 1637, nothing was further from his mind than a scheme to replace the construction-based conception of mathematics by one centered on formulas. On the

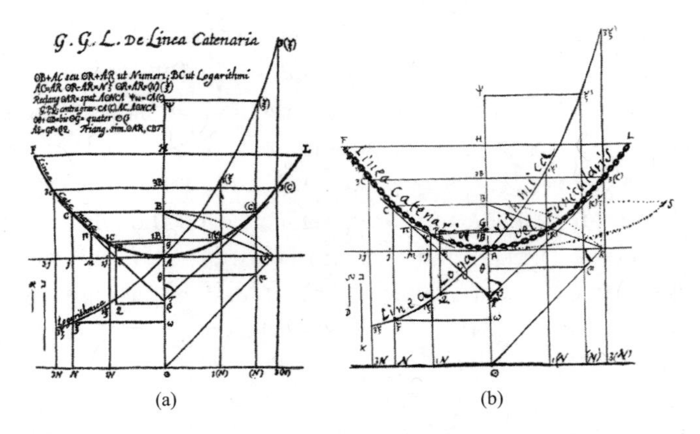

FIGURE 4. Leibniz's figures for the catenary, showing its relation to the exponential function. (From [5] and [8], respectively.)

contrary, his starting point was a new curve-tracing method, which he presented as a generalization of the ruler and compass of classical geometry, and he accepted algebraic curves only once he had established that they could be generated in this manner [3].

It is in this context that we must understand Leibniz's construction: He sees the catenary not as an applied problem to be reduced to mathematical formulas, but as a fundamental construction device analogous to the ruler and the compass of Euclidean geometry. (See Figure 4 for two of his original figures.) Extending the constructional toolbox with new curve-tracing devices along these lines was a major research program in the late seventeenth century. Beside the catenary, other physical curves were also called upon for this purpose, such as the elastica [1] and the tractrix [2].

Thus seventeenth-century mathematicians had reason to reject the "chicken scratch mathematics" that we take for granted today. They published not formulas but the concrete, constructional meaning that underlies them. If you want mathematics to be about something, then this is the only way that makes sense. It is prima facie absurd to define mathematics as a game of formulas and at the same time to assume naively a direct correspondence between its abstraction and the real world, such as $y = (e^x + e^{-x})/2$ with the catenary. It makes more sense to turn the tables: to define the abstract in terms of the concrete, the construct in terms of the construction, the exponential function in terms of the catenary. It was against this philosophical backdrop that Leibniz published his recipe for determining logarithms using the catenary. We see, therefore, that it was by no means a one-off quirk; rather it was a natural part of a concerted effort to safeguard meaning in mathematics.

References

1. V. Blåsjö, "The rectification of quadratures as a central foundational problem for the early Leibnizian calculus." *Historia Math.* **39** (2012) 405–431, http://dx.doi.org/10.1016/j.hm.2012.07.001.
2. H. J. M. Bos, "Tractional motion and the legitimation of transcendental curves." *Centaurus* **31** (1988) 9–62, http://dx.doi.org/10.1111/j.1600–0498.1988.tb00714.x.
3. ———, *Redefining Geometrical Exactness: Descartes' Transformation of the Early Modern Concept of Construction*. Springer, New York, 2001.
4. T. Hobbes, *The English Works of Thomas Hobbes of Malmesbury*. Vol. 7. Longman, Brown, Green, and Longmans, London, 1845.
5. G. W. Leibniz, "De linea in quam flexile se pondere proprio curvat, ejusque usu insigni ad inveniendas quotcunque medias proportionales & logarithmos." *Acta Eruditorum* **10** (1691) 277–281.
6. ———, "De solutionibus problematis catenarii vel funicularis in Actis Junii A. 1691, aliisque a Dn. I. B. propositis." *Acta Eruditorum* **10** (1691) 434–439.
7. ———, "De la chainette, ou solution d'un problème fameux proposé par Galilei, pour servir d'essai d'un nouvelle analise des infinis, avec son usage pour les logarithmes, & une application à l'avancement de la navigation." *J. Sçavans* (Mar. 1692) 147–153.
8. ———, "Solutio illustris problematis a Galilaeo primum propositi de figura chordae aut catenae ex duobus extremis pendentis, pro specimine nouae analyseos circa infinitum." *Giornale de' Letterati* (Apr. 1692) 128–132.
9. ———, *Über die Analysis des Unendlichen*. Ed., trans. G. Kowalewski. Engelmann, Leipzig, 1908.
10. ———, "Two papers on the catenary curve and logarithmic curve," trans. P. Beaudry, *Fidelio Mag.* **10** no. 1 (Spring 2001) 54–61, http://www.schillerinstitute.org/fid_97–01/011_catenary.html.
11. ———, *Sämtliche Schriften und Briefe. Reihe III: Mathematischer, naturwissenschaftlicher und technischer Briefwechsel. Band 5: 1691–1693*. Leibniz-Archiv, Hannover, Germany, 2003, http://www.leibniz-edition.de.

Rendering Pacioli's Rhombicuboctahedron

Carlo H. Séquin and Raymond Shiau

1. Introduction

The focus of attention in this article is the rhombicuboctahedron (RCO) (Figure 1) that appears in the painting "Ritratto di Fra' Luca Pacioli" (1495) exhibited in the Museo e Gallerie di Capodimonte [15] in Naples, Italy. The RCO is one of the 13 Archimedean solids. These are semiregular convex polyhedra composed of two or more types of regular polygons meeting in identical vertices; they are distinct from the Platonic solids, which are composed from only a single type of regular *n*-gon. The central character in Figure 1(a) is Fra' Luca Pacioli, a famous mathematician of the Renaissance period, most likely lecturing on some topic concerning the Platonic or Archimedean polyhedra. There is some speculation that the second person in the painting might be Albrecht Dürer, who also had an interest in symmetrical polyhedral objects. The great fascination with such objects at that time eventually culminated with *De Divina Proportione* written by Pacioli around 1497. This book on mathematical and artistic proportions was illustrated by Leonardo da Vinci.

The suspended RCO, shown in the top left of the painting, is composed of 18 squares and 8 equilateral triangles, and each vertex is shared by 3 squares and 1 triangle. The painting implies that this object has been realized with 26 glass plates fused together well enough that this shell can be filled with some liquid up to its centroid.

Mackinnon suggests [12, 13] that the RCO, with its square and triangular faces, represents four of the five regular Platonic solids, and that by filling it partially with water it might evoke associations with the four corresponding elements according to the chemical theory of Plato's *Timaeus*: *Earth* (cube) by the physical glass shell, *Water* (icosahedron), *Air* (octahedron) by the media contained in this shell, and *Fire* (tetrahedron) by the depicted bright reflections. The fifth regular solid, the dodecahedron, which for Plato represented the *Universe*, is shown in a model on the right-hand side of the painting.

This painting is now generally attributed to Jacopo de' Barbari. However, there is some speculation that the depiction of the transparent RCO was added by some other artist, since in style and detail it looks quite different from the other objects in this room. Many admirers of this painting have issued glowing comments about how wonderfully the painter seems to have captured the reflections and refractions in this object (Figure 1(b)). Mackinnon [13] attributes this part of the painting to Leonardo da Vinci himself. Even though Pacioli had not yet met Leonardo when this panel was first painted, and he started to collaborate with him only after 1495, there is a possibility that the RCO was added later to the portrait of Pacioli when they were both in Milan [12]. Many conjectures and some speculation have originated from this painting, and the discussion of its creation and history is still ongoing [2, 5, 8, 19].

Clearly, the RCO in this painting stands out in a special way. On his web page devoted to Luca Pacioli's Polyhedra, Hart states [10]:

> The polyhedron in the painting is a masterpiece of reflection, refraction, and perspective. (Davis states that the bright region on its surface reflects a view out an open window, showing the Palazzo Ducale in Urbino.) Certainly an actual glass polyhedron was used as a model. (Pacioli states in his books that he constructed several sets of glass polyhedra, but I know of no other information about them.) The polyhedron in the painting is beautifully positioned,

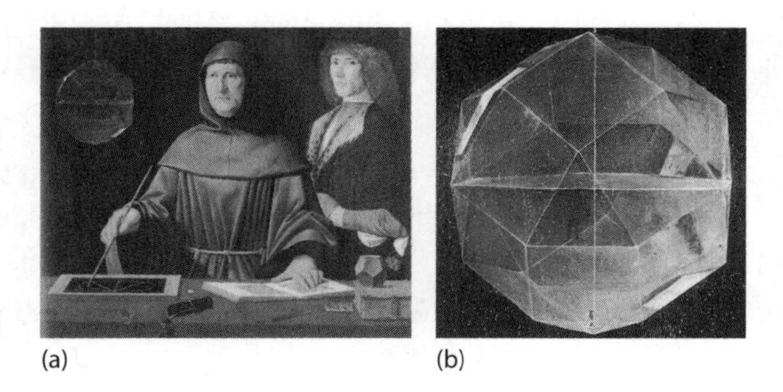

FIGURE 1. (a) Pacioli painting; (b) enlarged and enhanced view of the rhombicuboctahedron (RCO). See also color images.

suspended with a 3-fold axis vertical, out of physical contact with the other objects in the scene. I suspect that Pacioli chose it for the portrait because he discovered this form and was quite proud of it. (Presumably Archimedes first discovered it, but that wasn't known in Pacioli's time.) The painting is the earliest known image of the rhombicuboctahedron.

However, on a closer, more critical inspection, some things seem not quite right: the reflection of the window (not seen in the painting) on the upper left of the RCO seems more like a pasted-on sticker image that bends around one of the polyhedron edges than two separate reflections in the two differently angled RCO facets. In 2007, Rekveld raised doubts in a publication [17] and on his website [18]:

> Last week in Napoli I revisited the Capodimonte museum and its amazing collection of paintings. At some point I found myself face to face with this canvas, attributed to Jacobo de Barbari, a portrait of the mathematician Luca Pacioli painted in 1495.
>
> . . . In said painting I suddenly noticed the mysterious reflections in the gorgeous mathematical shape at the top left, a rhombicuboctahedron, made of glass and half filled with water. I took a picture, and when I zoom in on these painted reflections we see buildings and sky, as if to suggest that an open window in the room is being reflected in the glass facets. The direction doesn't seem right though; am I wrong or do the reflections show that the window is high up towards the left? Or rather towards the bottom right? Both do not really make sense. Also in either case the light in the painting does not seem to be much affected by the source of these reflections.
>
> . . . To me the way the same window seems to be reflected three times within this shape doesn't seem terribly true to the laws of optics. Or more precisely: they seem true to a textbook notion of reflection and refraction of light in water, but the way the images are formed doesn't seem very realistic at all.
>
> . . . Its depiction doesn't seem based too much on observation or the tracing of reality through a camera obscura, but then again, I've never actually seen a glass rhombicuboctahedron half filled with water, so who am I to judge?

Thus, it seems worthwhile to investigate these issues. Just as disturbing as the "pasted-on" reflection of the window discussed above is the fact that there are no visible effects of refraction in the water body in the lower half! The RCO edges on the back surface appear in the painting in exactly the places where the computer rendering of a thin wire-frame object (Figure 2(b)) shows them.

In May 2011, Claude Boehringer [4], an artist working with various materials, produced a physical glass model of an RCO shell held together with lead, which was strong enough to be half filled with water. Under the guidance of Herman Serras [22], the model was suspended in the proper way to obtain the same orientation as the polyhedron in the painting. Serras then took the photograph shown in Figure 2(a). This image looks quite different from the one in the painting, and so it is difficult to draw conclusions about the realism of the depiction of 1495. The surroundings where this model was photographed are entirely different, and there are no dominant reflections of a single, brightly lit window. Nevertheless, the photo of the Boehringer model (Figure 2(a)) confirms our intuition that those edges seen through the water body would be seriously altered in their rendered positions.

In 2011, a discussion arose concerning perceived geometrical flaws in some of Leonardo's drawings [11]. Huylebrouck's note also made a reference to the RCO in the portrait of Pacioli, and this caused one of the authors to take a closer look at the depiction of this object [20]. While there were no geometrical errors in the projection of this object, some of the visual effects due to reflection and refraction seemed clearly wrong. This raised the question about what an actual RCO glass shell half filled with water really would look like. Several students at Berkeley and elsewhere responded to the challenge, using

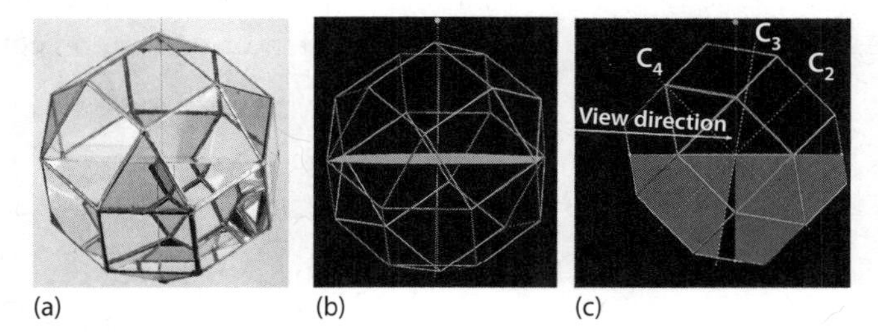

FIGURE 2. (a) Model by Claude Boehringer photographed by Herman Serras; (b, c) viewing geometry. See also color images.

computer graphics programs to model such an object with its various reflections and refractive effects. However, the results offered looked quite different, driving home the point that modeling a complex object with multiple volumes with different refractive indices abutting one another is not a trivial task. When making a model of the object to be rendered, the idiosyncrasies of the rendering program to be used need to be taken into account carefully. One model does not fit all possible renderers! Even for people experienced in using computer graphics rendering programs, it is advisable to run through a series of progressively more complex test models when switching to a new renderer.

In most of our efforts, we have used Autodesk *Maya* [1] as our modeling tool and *Mental Ray* [14] as our rendering engine using a basic ray-tracing algorithm [9, 16]. First, we constructed simple geometrical test models for which the correctness of the renderings could readily be verified and made sure that the renderer properly interpreted the desired geometry in the various refraction events and reflections at external and internal boundaries. In the end, we repeated this process once more for the open-source rendering program *Blender* [3]. Overall, these computer simulations confirm that it is highly unlikely that the painter was observing a physical glass container half filled with a clear liquid when painting the RCO.

In the following sections, we briefly discuss the geometrical set-up of the RCO for a geometrical analysis in the context of the article as well as for our computer modeling. A detailed discussion of our series of test renderings, which we recommend as a preparatory debugging step for challenging rendering tasks, can be found in a technical report [21].

Our computer simulations look quite different from the painted RCO, proving that the latter is not a physically correct rendering. We thus conclude that the artist's most likely objective was to create the most "plausible" and "convincing" depiction of such an object for a broader public.

2. *Viewing Geometry*

To create a realistic rendering that can be compared with the painted RCO, we first have to figure out how this RCO has been suspended and where to place the observer's eye. If the RCO were indeed suspended

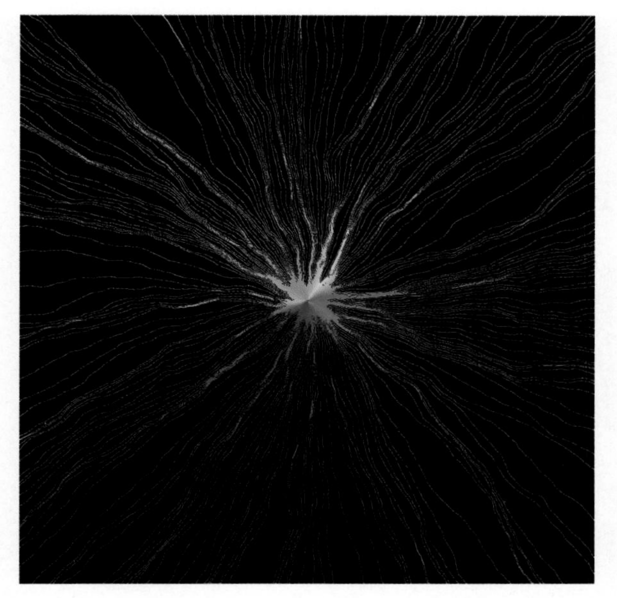

FIGURE 3 from "A Unified Theory of Randomness." Photo by Scott Sheffield.

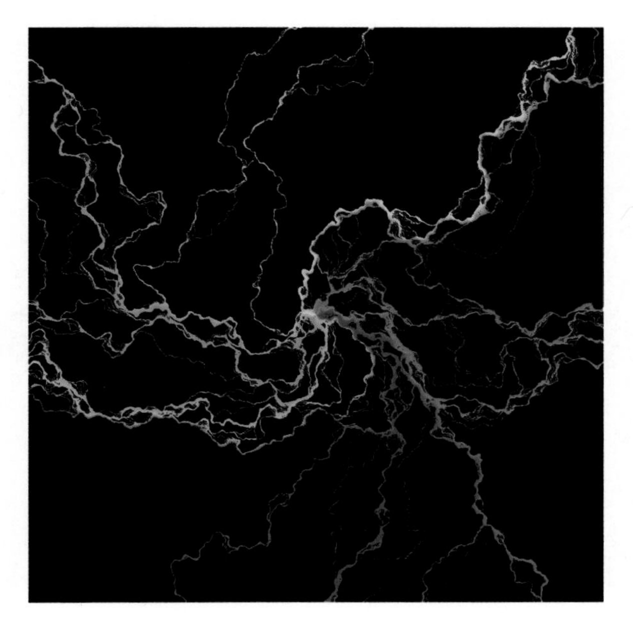

FIGURE 4 from "A Unified Theory of Randomness." Photo by Scott Sheffield.

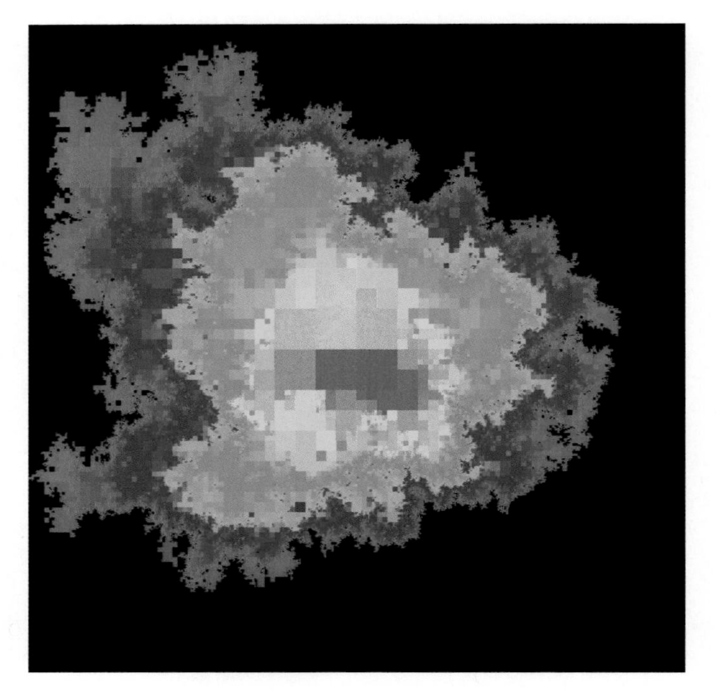

FIGURE 8 from "A Unified Theory of Randomness." Photo by Jason Miller.

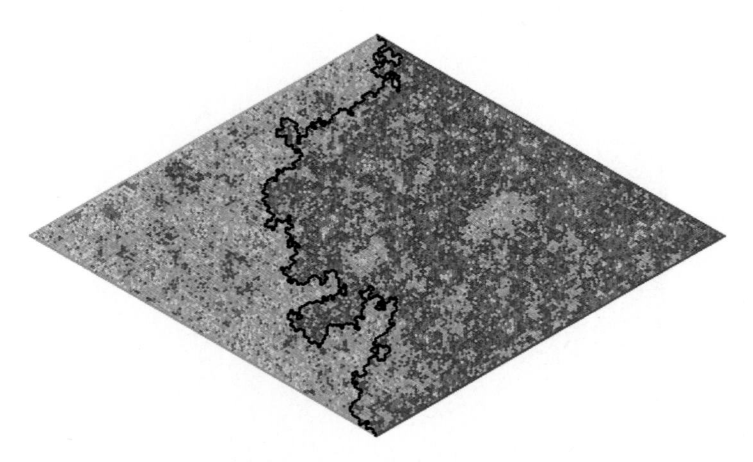

FIGURE 11 from "A Unified Theory of Randomness." Photo by Jason Miller.

Figure 1a from “Creating Symmetric Fractals.”

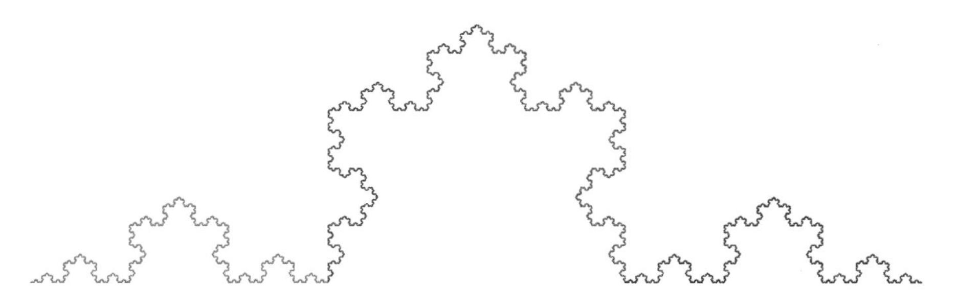

Figure 1b from “Creating Symmetric Fractals.”

FIGURE 1c from "Creating Symmetric Fractals."

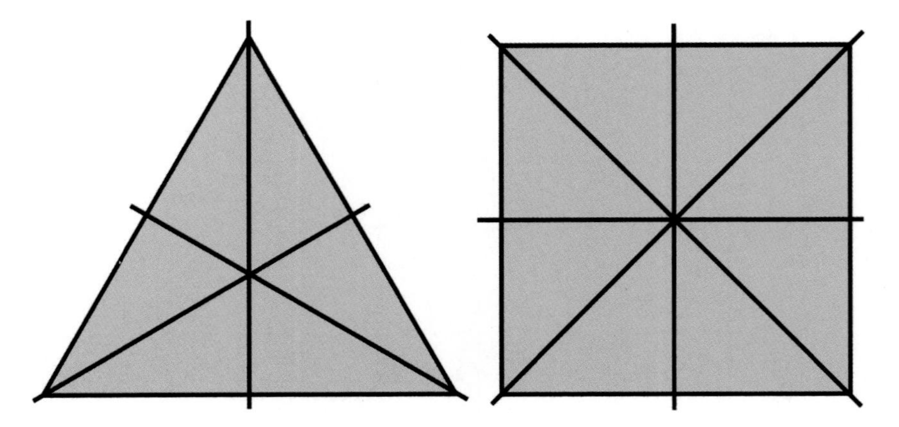

FIGURE 2 from "Creating Symmetric Fractals."

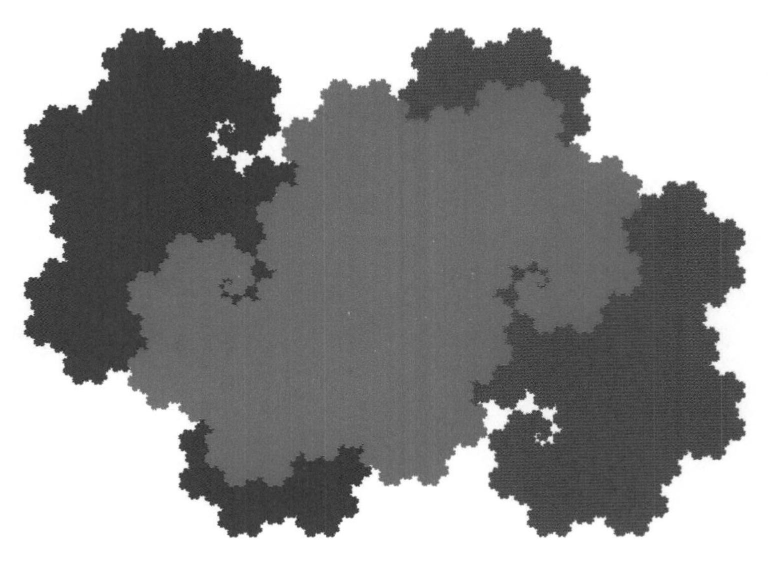

FIGURE 3 from "Creating Symmetric Fractals."

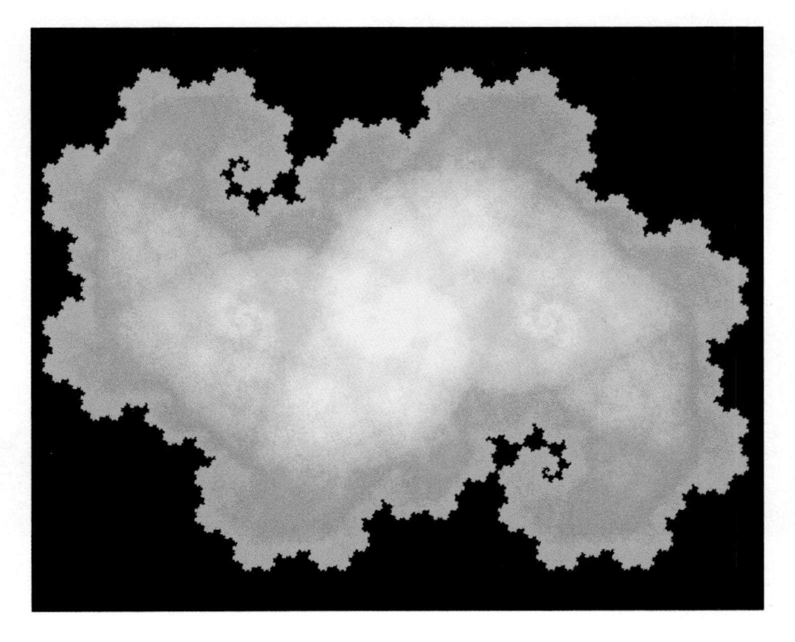

FIGURE 4 from "Creating Symmetric Fractals."

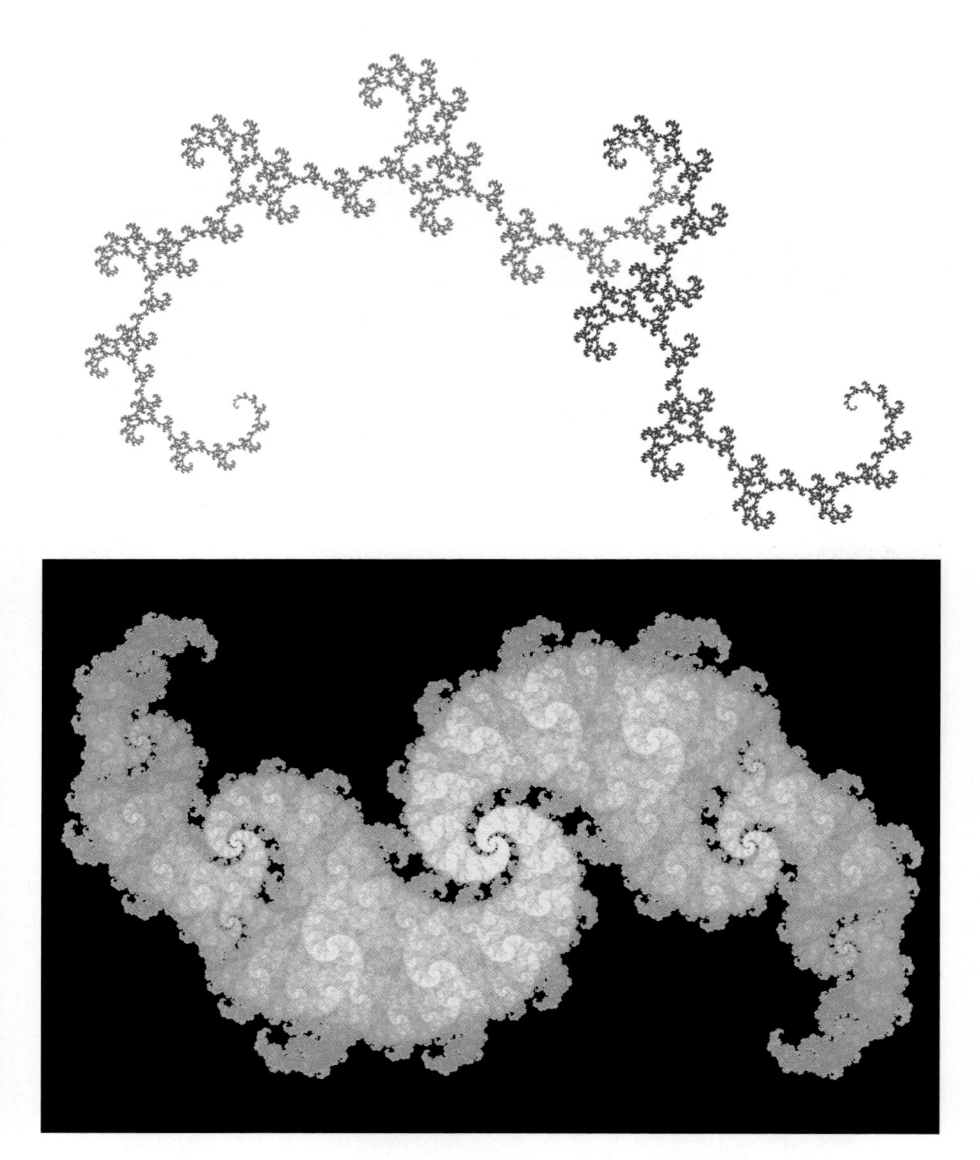

Figure 5 from "Creating Symmetric Fractals."

Figure 6 from "Creating Symmetric Fractals."

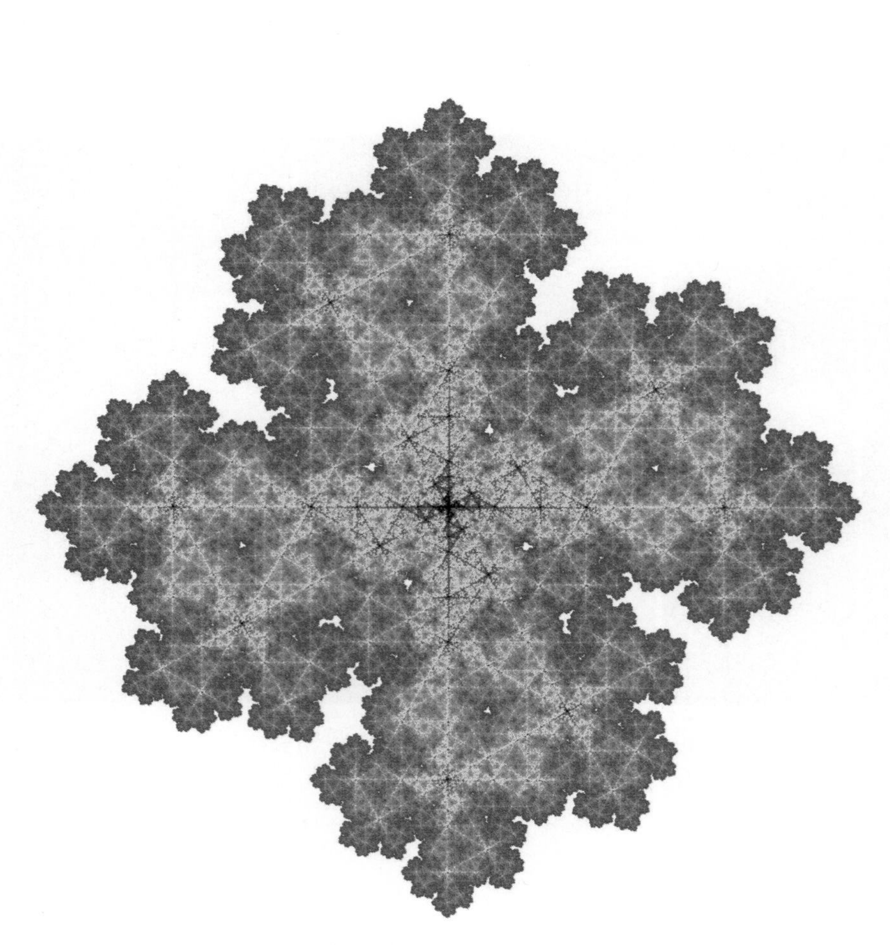

FIGURE 7a from "Creating Symmetric Fractals."

Figure 7b from "Creating Symmetric Fractals."

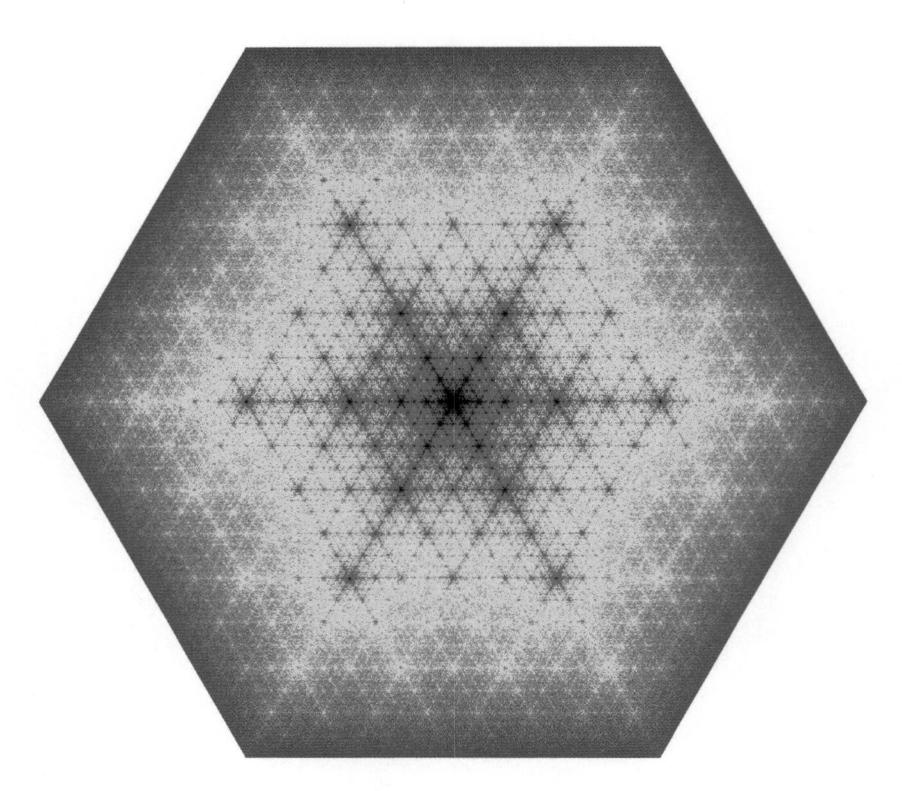

FIGURE 8a from "Creating Symmetric Fractals."

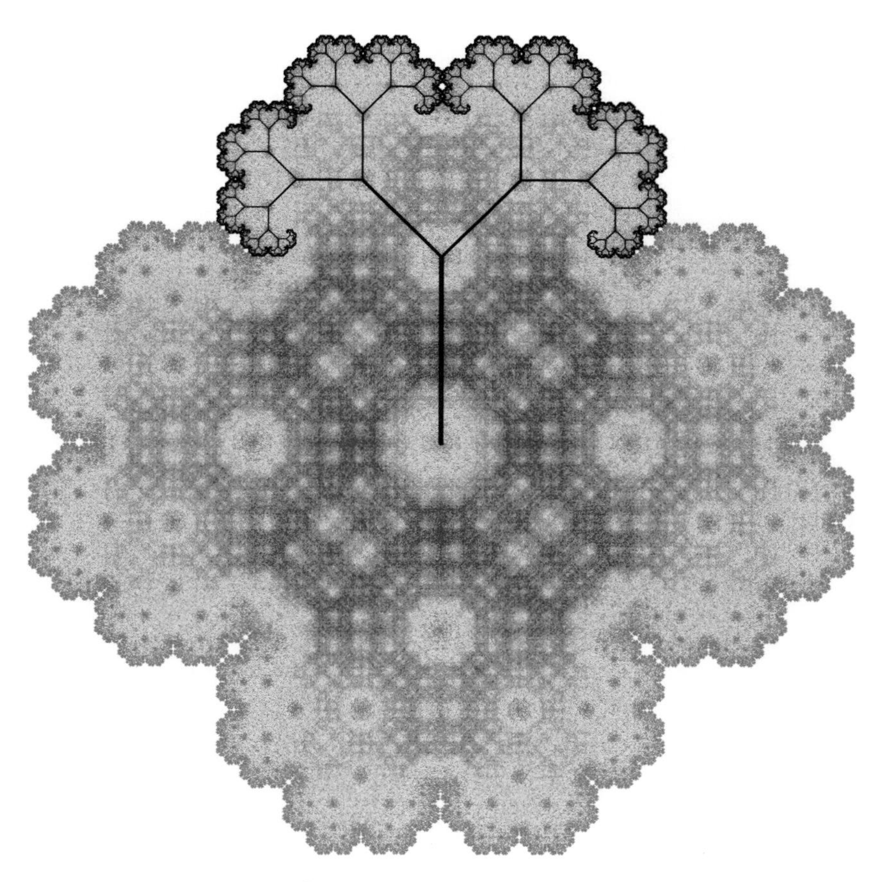

FIGURE 9 from "Creating Symmetric Fractals."

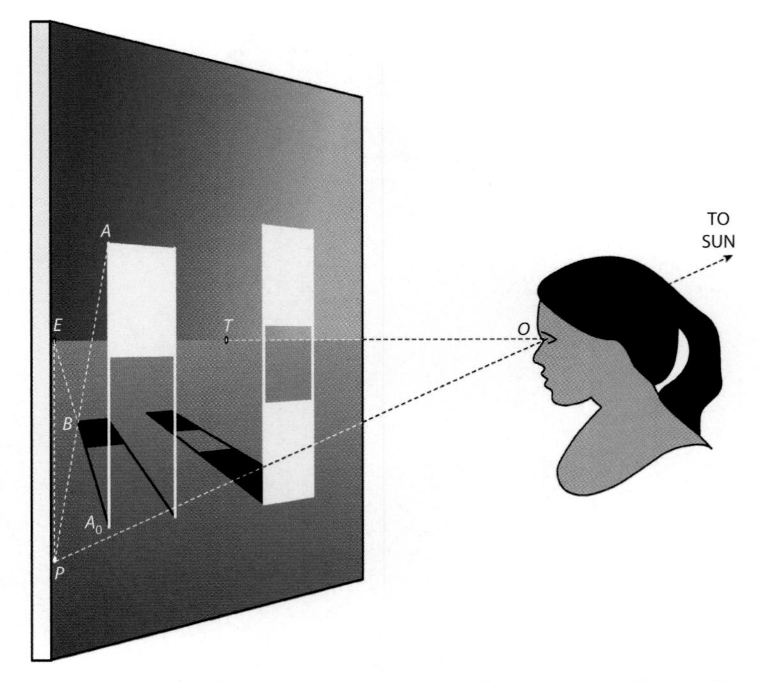

FIGURE 7 from "Projective Geometry in the Moon Tilt Illusion."

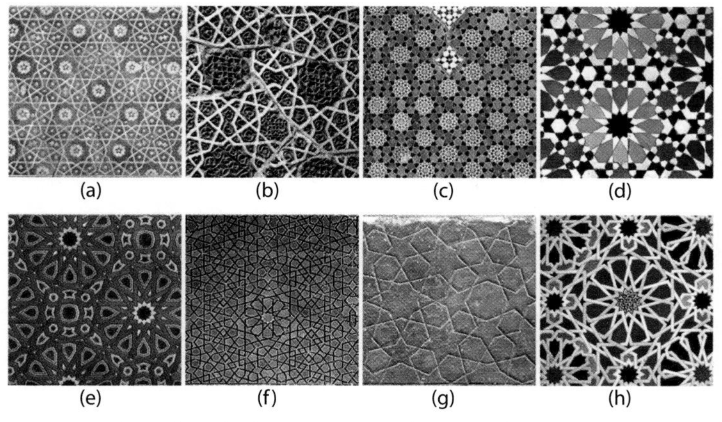

Figure 1 from "Girih for Domes: Analysis of Three Iranian Domes."
(Pattern in Islamic Art 2015)

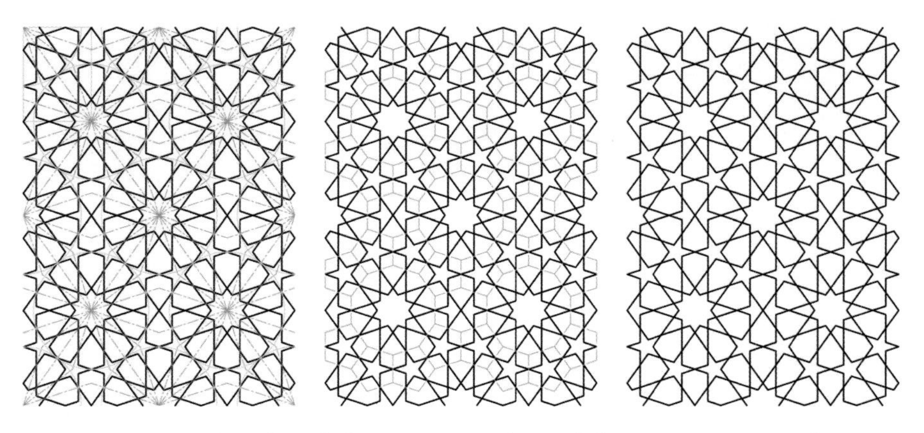

Figure 3 from "Girih for Domes: Analysis of Three Iranian Domes."

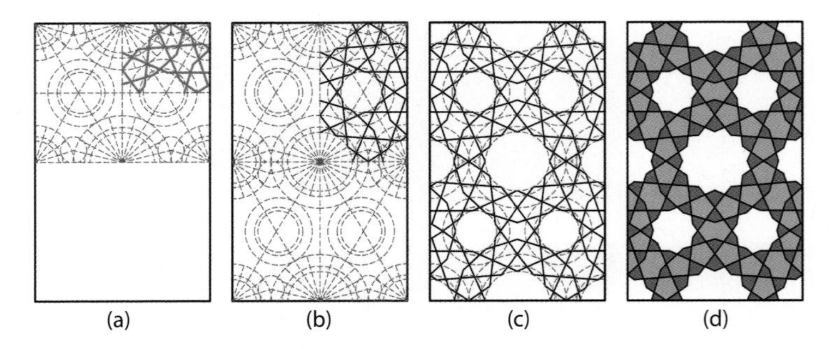

FIGURE 5 from "Girih for Domes: Analysis of Three Iranian Domes." Designed by Hussein Lorzadeh (Ra'eesZadeh and Mofid 2011:181).

FIGURE 7 from "Girih for Domes: Analysis of Three Iranian Domes."

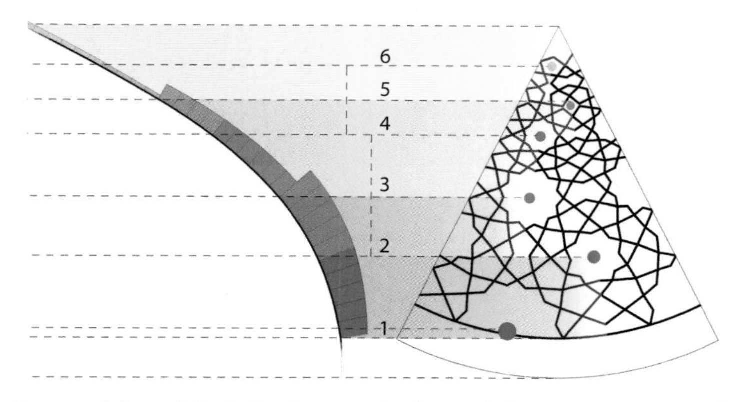

FIGURE 9 from "Girih for Domes: Analysis of Three Iranian Domes."

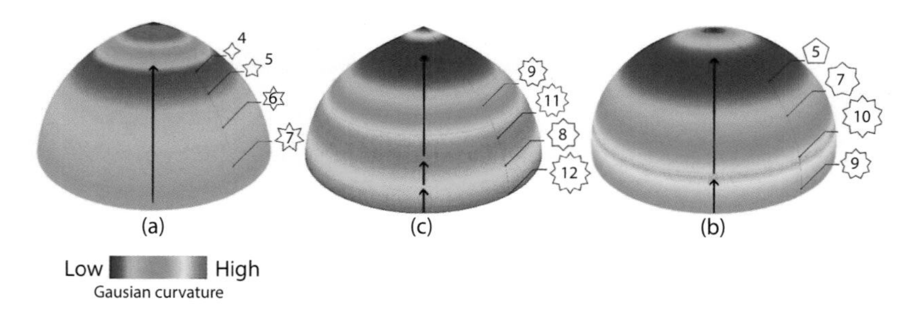

FIGURE 10 from "Girih for Domes: Analysis of Three Iranian Domes."

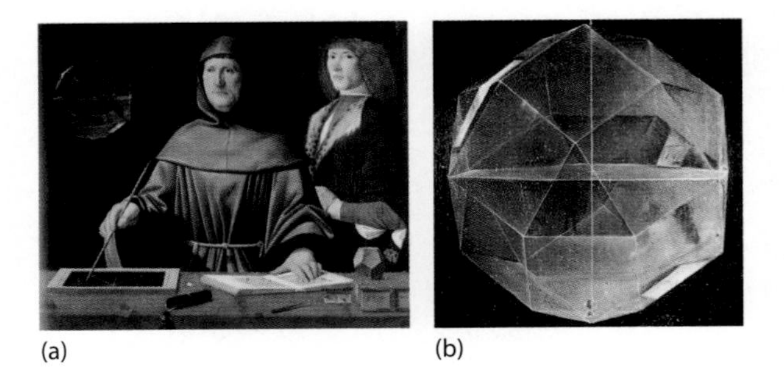

FIGURE 1 from "Rendering Pacioli's Rhombicuboctahedron."

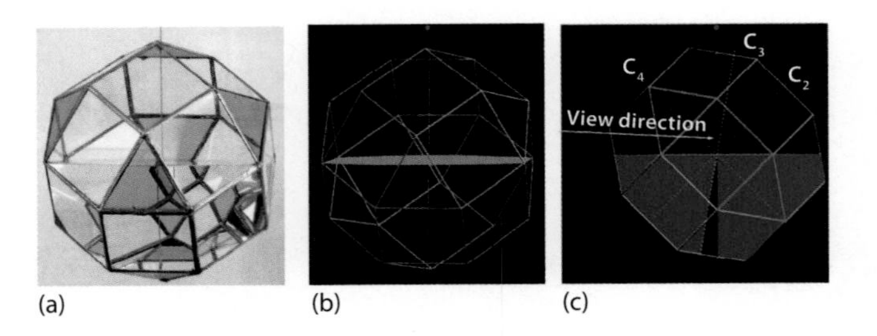

FIGURE 2 from "Rendering Pacioli's Rhombicuboctahedron."
Part (a) model by Claude Boehringer photograped by Herman Serras

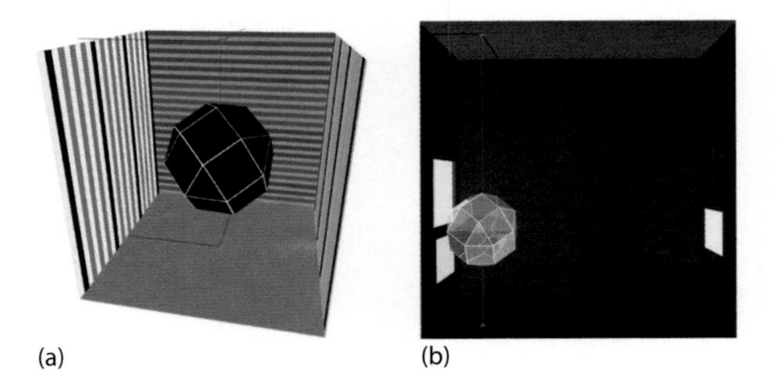

FIGURE 4 from "Rendering Pacioli's Rhombicuboctahedron."

along one of its three-fold symmetry axes piercing the centers of the top and bottom triangles, as stated by Hart [10], then these two faces would be truly horizontal. In the painting, we see both of these triangles from below, which would then imply that the eye of the observer must lie below the bottom of the RCO. On the other hand, the horizontal water surface clearly is seen from above: the more strongly depicted edges of the RCO belong to its front facets, and the foreshortened left and right edges of the polygon depicting the water surface have a vanishing point that lies somewhere in the upper right. This view geometry requires an eye point above the center of the RCO.

Careful inspection of the enlarged view of the painted RCO (Figure 1(b)) shows that the water surface passes through 4 of the 24 vertices of the RCO and cuts the vertical, square facets on the left and right sides along their horizontal diagonals (Figure 2(b) and 2(c)). This implies that the suspension line must form an angle of 45° with the plane of the square immediately in the back of the top triangle, as well as with the square in front of the top of the RCO. The C_3 symmetry axis deviates by $54.74° - 45° = 9.74°$ from this suspension line. A few trigonometric calculations then reveal that this suspension line cuts the altitude of the top triangle in the ratio $1{:}\sqrt{2}$, i.e., a fraction of 0.4142 away from the top vertex in the painting. Serras [22] also had calculated the distance of the suspension line penetration from the closest triangle vertex as $(\sqrt{3})(\sqrt{2} - 1)/2 = 0.3587$ times the RCO edge length. Fortunately, these two calculations agree. In addition, the water boundary intersects two more pairs of triangle edges at a fraction of $\sqrt{2} - 1 = 0.4142$ away from the shared vertex (highlighted dots on the left and right edges at the water's surface in Figure 2(c)).

With the angle of suspension unambiguously resolved based on the boundary of the water level, we now can try to locate the eye point of the observer. The viewer must be looking toward the center of the RCO with a slight downward angle that must lie between 0° and 9.74°. Using an interactive computer graphics program that let us readily adjust the viewing parameters, we found a best match between an appropriately tilted computer model and the rendering in the painting for a downward angle of about 3° (Figure 2(c)). Comparing the resulting view of a wireframe RCO (Figure 2(b)) with the polygon edges depicted in the painting (Figure 1(b)) confirms that we have found a suitable approximation of the correct viewing geometry.

Now let us review the above findings in the context of the complete painting. First, we try to find the viewer's eye level with respect to the painting. Our intuition tells us that the observer's eyes are at about the height of Pacioli's nose or eyes and that the person on the right is looking slightly down on the viewer. This point of view would place the view center properly above the water level. We can also try to find a horizon compatible with the objects on the table (assumed to be horizontal), and this is what we found:

(1) a vanishing point on top of Pacioli's head for the slate tablet;
(2) a vanishing point somewhat above the middle of the RCO (but far out to the left) for the open book (with substantial error margins); and
(3) a vanishing point slightly above the top of Pacioli's head for the red box (also with substantial error margins).

All of these vanishing points are higher than the center of the RCO, and this is compatible with our view of the water surface from above. But then, the fact that we can see the lower side of the bottom triangle implies that it must be slanted upward towards the viewer. This confirms the tilted suspension of the RCO established above (Figure 2(c)).

Now let us take a closer look at the projection used in the painting. Clearly, there is some perspective involved; the back-face triangle (pointing down) is a few percent smaller than the size of the front triangle (pointing up). The same interactive computer rendering program that lets us find the best match for the view angles also lets us find the parameters for the perspective projection in which the rendered edge crossings of the wire frame model match the locations in the painting as closely as possible; this occurs when the eye is about 14 RCO diameters away from its center. Now, based on its position and comparing it to the hands and head of Pacioli, we may estimate that the RCO is about 8–10 inches in diameter. Thus, based on its own perspective, it must have been drawn as it would look from about 10–11 feet away. However, Pacioli in the painting seems only about 5–6 feet away from the viewer, and the object, if it is indeed above the table, would then be only 4–5 feet away. Thus, the perspective of the RCO is not compatible with the rest of the scene. The RCO should exhibit much stronger foreshortening, or it should have been depicted at only about half its current scale. This raises a strong suspicion that this RCO was sketched out quite carefully, but separately—perhaps from a smaller model or from a geometrical construction—before it was copied into the Pacioli painting.

3. Computer Modeling and Rendering of the RCO

Let us assume that the RCO was painted in a separate sitting, possibly in a room with an open window in the left wall, offering a view of a palazzo and some bright sky, and perhaps with some more local illumination coming from the lower right. Can we find an environment and some suitable material constants and illumination levels that will produce an image closely resembling the depiction in the painting? What would a physically more realistic rendering of such a glass container half filled with water look like?

The first task is to create an appropriate geometrical model of the object to be rendered. To have some control over the appearance of the edges of the glass container, we flesh out the wireframe shown in Figure 2(b) into properly mitered prismatic beams that are composed into a polyhedral object of genus 25 (i.e., a shell with 26 facet openings). Figure 3(a) shows an enlarged partial view of this framework; to make its geometry more clearly visible, all the cross sections have been enlarged by a factor of 2. In our model, the thickness of these struts is parameterized and set equal to the thickness of the glass plates shown labeled as GLASS in Figure 3(b); this allows us to emphasize or deemphasize the visibility of these seams. In most of our renderings, we have set the material of the struts to be some grey, diffusely reflective material. However, there is the option to assign them the same material parameters as are used for the glass plates, in order to simulate a completely fused contiguous glass container. Finally, the water volume is modeled as a separate polyhedron that fits snugly against the inside surfaces of the glass plates in the lower half of the RCO (Figure 3(c)).

Once we have gained confidence that the chosen modeling technique and the adopted rendering environment produce physically justifiable results [21], we can model the RCO as described in Figure 3 and render it in various environments using a basic ray-tracing program [3, 14]. Overall, we have three different transparent materials (Figure 3(b)): air (A), water (W), and glass (G), plus the opaque surfaces (O) associated with the filler framework located around the edges of the glass plates.

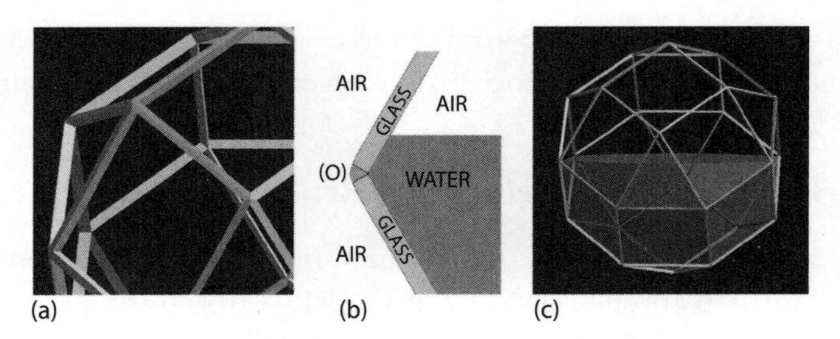

FIGURE 3. Modeling the complete RCO: (a) filler framework; (b) cross section showing the types of interfaces; (c) a transparent, nonrefractive water polyhedron inside the opaque filler framework.

Any ray that hits the latter type of surface is terminated and returns the illumination value found at this location.

The interfaces between two transparent materials are modeled as smooth surfaces that permit refraction as well as reflection processes to occur. In general, they will divide an incoming ray of generation n into two subrays of generation $n + 1$; the relative strengths of the two subrays are determined by the Fresnel equations [7]. Special care must be taken to create the properly merged interfaces where the body of water is in contact with the glass plates forming the container. For 8 of the 26 glass plates, their inner surfaces had to be split into two parts—one interfacing with water and the other interfacing with air. In the complete model, there is one interface facet of water/air, 26 + 17 instances of glass/air and 17 instances of glass/water.

We started with some extensive testing of the RCO model [21] in the synthetic environment of a virtual Cornell Box [6] with distinctly striped walls (Figure 4(a)), where the final stopping point of any ray being reflected on or refracted through the RCO can be identified more easily. Those tests gave us the confidence that our rendering process will give us a depiction that is close to physical truth with respect to the placement of the various visual elements resulting from reflection and refraction.

To produce the final renderings (Figure 5), we have suspended the RCO in a darkened room with three brightly colored "windows" (Figure 4(b)); these are really just bright, flat "paintings" on the walls. The RCO is placed close enough to the large cyan "window" on the left wall, so that portions of it appear in both the upper left panels of the

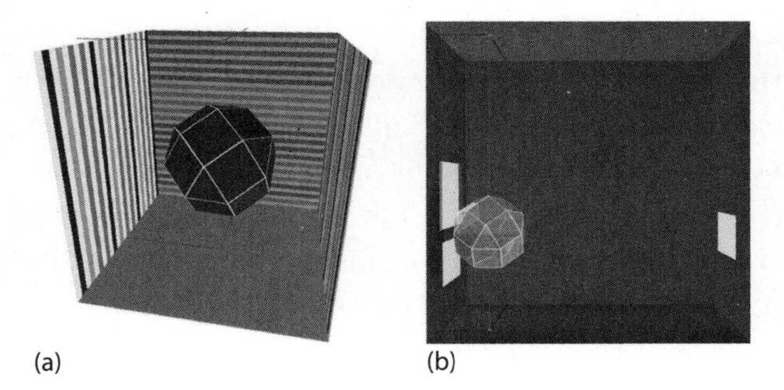

FIGURE 4. Placement of the RCO inside virtual Cornell Boxes: (a) for detailed studies of refractions and internal reflections [21], and (b) for final studies of external window reflections. See also color images.

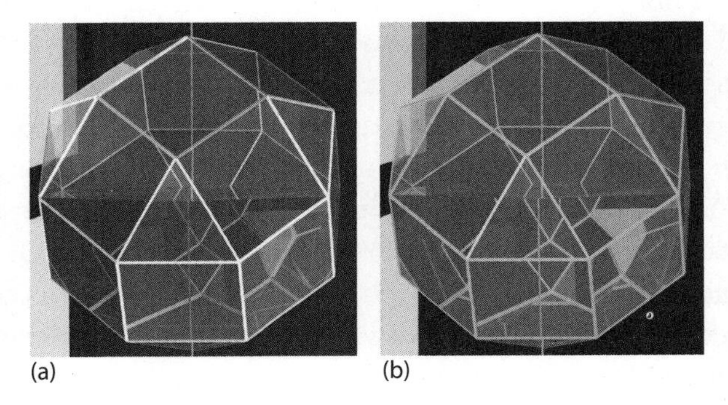

FIGURE 5. RCO in a dark room with windows: (a) rendered with *Maya* to a recursion depth of 10; (c) rendered with *Blender* to a depth of 40.

RCO, where the original painting shows the primary window reflections. This "window" is enhanced with blue and yellow vertical edges to make it easier to understand what is being reflected in those two facets of the RCO. One can see the blue back edge of that window in the upper panel; the yellow front edge appears in the lower facet. This disparity in reflected window portions illustrates the physical impossibility of a single continuous image of the window appearing across both faces, as depicted in the Pacioli painting.

There are other features on the surface of the painted RCO that can be interpreted as depicting reflections. The lower right of the RCO has some pronounced greenish tint that can be seen as a reflection of the green tablecloth and a fuzzy dark feature might represent one of the bodies standing behind the table. More distinctly drawn, a small black reflection in the front triangular face of the RCO (Figure 1(b)) was probably meant to represent the artist who drew the image of the RCO; however, it is unrealistically small. A human observer standing some distance *d* in front of a plane mirror will see his reflection as being located at that same distance *d* behind the mirror, and the perceived 'image' at the reflecting surface will be half the size of the observer. Thus, the black figure outline is much too small to be a reflection of the artist; to a first approximation, it should be about the same size as Pacioli or his companion. In our computer rendering, we created a comparable black reflection by placing a six-inch-tall puppet cutout in the plane of the camera. Moreover, since the triangular front face of the RCO is not perpendicular to the line of sight, the puppet had to be displaced about 10 body lengths downward and to the left of the position of the camera lens—supposedly the eye of the artist.

The computer renderings (Figure 5) also show internal reflections in quite different locations compared with the painted RCO. For instance, we see portions of the light blue window and its dark blue back edge in two facets in the bottom right quadrant of the ray-traced RCO, but none of these have a similar appearance in the painted RCO, even accounting for the murkiness of the painted water. Conversely, the painted RCO shows some internal reflections that are not seen anywhere in the ray-traced RCO: in the back/right center quad facet above the water surface, and along the bottom/right of the RCO. In attempting to replicate these extra internal reflections seen in the painted RCO, we expanded the windows to fill the entire left wall, but still we could not obtain the desired internal reflection in the upper half of the RCO.

As mentioned before, the refracted locations of the string and the filler framework also appear in inappropriate locations. In the painting, the string appears as a continuous line above and below the water surface, whereas in the computer rendering, the string is discontinuous at the water surface. The displacement is caused by the different indices of refraction of air and water. The same applies to the appearance of the filler framework; our ray-traced RCO renders the refracted framework struts in very similar locations to the physical model shown in Figure 2(a)—in contrast to the painted RCO.

In order to check the reproducibility of our rendering results and the robustness of the step-by-step validation approach recommended above, we also experimented with a different rendering program: *Blender* [3] is an open-source program readily available to everybody and may therefore be a candidate that many readers might use for quick experiments. Once our model data had been adjusted to be compatible with the *Blender* environment, the rendered geometrical features looked the same as they look with the *Maya* renderer. We could also confirm that increasing the depth of the recursive ray tree from 10 to 40 did not result in any meaningful differences; any changes that might be attributable to increased ray-tree depth are small compared to the noticeable variations in the displayed intensities of reflected and refracted components between the two rendering programs. The major differences between the two renderings occurred with respect to hue changes and brightness variations. Different rendering programs use different representations for the relative importance of the specularly reflected, refracted, and diffusely reflected photons, resulting in different relative intensities of some of the struts as seen through multiple layers of reflection and refraction.

More troublesome for our particular rendering comparison was the fact that *Blender* permits volume tinting only for domains surrounded by a closed manifold surface, but not for volumes that have some individually merged interfaces as part of their boundary representation. Thus, the tinting of the glass and water components had to be achieved entirely via some approximate changes in surface parameters. This explains why the reflections of the pink window are much more vivid in the *Blender* rendering: the rays experience less tinting while traveling a substantial distance through water. This emphasizes yet again the need to carefully evaluate the capabilities and idiosyncrasies of any rendering environment, using an incremental approach that starts with some simple, well-understood test geometries.

There are still many parameters that could be fine-tuned: the refractive index of glass; the amount of tinting imparted by the glass plates and the water body; and the amount of diffuse scattering on the outer surfaces, perhaps caused by the presence of dust. We chose some of these parameters to make our rendering look somewhat similar to the depiction in the Pacioli painting, as depicted in Figure 1(b). However, our main focus was to show the geometrical issues related to the placement of the reflected and refracted geometrical features, because these effects are most relevant to the question of whether the artist was actually observing an RCO half filled with water.

4. Discussion and Conclusions

The RCO plays an important role in the Capodimonte canvas. It depicts an important accomplishment of the famous mathematician Fra' Luca Pacioli and was clearly meant to impress viewers of this painting. Baldasso [2] even argues that the RCO plays the role of a third figure in this composition.

Our analysis indicates that it is highly unlikely that the artist who painted the RCO in the Pacioli painting was looking at a physical glass RCO partly filled with water of the size implied in this painting. One of the anonymous reviewers contributed the following valuable information:

> In 1495 it was possible to make a small glass model of a polyhedron, and several are mentioned in two sixteenth century inventories of the Ducal palace at Urbino where Pacioli had worked; for technical reasons to do with the available glass it is not possible that these models were as large as the Pacioli RCO appears, nor could they have been strong enough to hold several liters of water.

This provides strong support for our claim that the artist did not directly observe the depicted object. However, it seems quite likely that the artist used an empty RCO model made of triangular and square glass plates for generating an original sketch, because the depicted geometry of the RCO edges is in excellent agreement (including perspective distortion) with the computer-generated rendering. When this initial depiction was later copied into the Pacioli painting, the discrepancies in the RCO's scale and its perspective projections, discussed at the beginning of this article, were introduced.

The primary window reflection across the top left facets is so far off from any possible reality that it cannot have been drawn based on an actual observation. It may, however, be possible that the painter observed the general nature of some window reflections on various polyhedral glass models and decided that reality was much too confusing for most observers to yield a pleasing painting. The artist may then have made a conscious decision to render a window reflection that would be more plausible and would help to identify the location of the space portrayed. A contiguous picture of the "Duomo in Urbino" can readily be understood by most viewers, whereas physically realistic reflections that break this image into disconnected pieces would be puzzling to most observers.

Moreover, to emphasize the transparency and three-dimensionality of the RCO, two other internal reflections of the same window were added in two facets on the right-hand side, one just above the water surface, and one on the bottom right in the water. The first one occurs where one would expect to see a secondary image, if this were the result of an intersection of a parallel beam of light coming through the assumed window. Furthermore, the contrast and saturation of these imagined reflections appear unnaturally strong compared to the other depicted features in the RCO. In addition, the apparent luminosity of the three reflections has been enhanced by the artist, by surrounding them with a slightly darkened halo.

One reason why this rendering of the RCO looks so good to an uncritical observer is that our visual system is much less sensitive to the precise effects of refractions and reflections, compared to shadows or perspective. It is plausible to assume that this is a consequence of our evolution during the last million years, when humans were trying to survive in the natural world, where reflection and refraction occur only in a rather limited way.

Though we have focused in this article on the defects with respect to a physically realistic rendering of the RCO, we should remember that physical realism was not the primary objective of the painter. A more likely goal was to present this intriguing mathematical object in the best possible way that would bring about a sense of wonder and awe in most viewers. We agree with many art historians that this goal has been achieved extremely well despite the discussed rendering flaws.

Acknowledgments

We would like to thank the anonymous reviewers for their many constructive comments and for some valuable insights into the context in which the discussed painting had been created.

References

[1] Autodesk *Maya*, software available at http://www.autodesk.com/products/autodesk-maya/overview. May 2017.

[2] R. Baldasso, "Portrait of Luca Pacioli and Disciple: A New, Mathematical Look," *The Art Bulletin*, March–June 2010.

[3] *Blender*. Blender Foundation; software available at http://www.blender.org/. May 2017.

[4] C. Boehringer, *Physical model of the RCO*, private communication. claude.boehringer@mac.com.

[5] A. Ciocci, *Il Doppio Ritratto del Poliedrico Luca Pacioli*. De computis. *Revista Española de Historia de la Contabilidad* 15 (2011), pp. 107–130.

[6] The Cornell Box. Cornell University; software available at http://www.graphics.cornell.edu/online/box/. May 2017.

[7] *Fresnel's Equations*. Available at http://hyperphysics.phy-astr.gsu.edu/hbase/phyopt/freseq.html. May 2017.

[8] E. Gamba, *L'Umanesimo matematico a Urbino*, in *La città ideale: l'utopia del Rinascimento a Urbino tra Piero della Francesca e Raffaello*, A. Marchi and M. R. Valazzi, eds., Electa, Milan, 2012, pp. 233–247.

[9] A. S. Glassner, *An Introduction to Ray Tracing*, Morgan Kaufmann, 1989.

[10] G. Hart, *Luca Pacioli's Polyhedra*. Available at http://www.georgehart.com/virtual-polyhedra/pacioli.html. May 2017.

[11] D. Huylebrouck, *Lost in Triangulation: Leonardo da Vinci's Mathematical Slip-Up*. (posted March 29, 2011) Available at https://www.scientificamerican.com/article/davinci-mathematical-slip-up/. May 2017.

[12] J. Logan, *An Analysis of Certain Mathematical Cyphers in the Portrait of Fra Luca Pacioli*, Unpublished draft; private communication, January 19, 2015.

[13] N. Mackinnon, "The portrait of Fra Luca Pacioli." *Math. Gazette*. 77 (1993), pp. 130–219.

[14] *Mental Ray* Documentation. Available at http://docs.autodesk.com/MENTALRAY/2012/CHS/mental%20ray%203.9%20Help/files/tutorials/architectural-library.pdf. May 2017.

[15] Museo e Gallerie di Capodimonte, Ritratto di Fra' Luca Pacioli. Available at http://cir.campania.beniculturali.it/museodicapodimonte/itinerari-tematici/galleria-di-immagini/OA900154. May 2017.

[16] G. S. Owen, *Ray Tracing*. Available at https://www.siggraph.org/education/materials/HyperGraph/raytrace/rtrace0.htm. May 2017.

[17] J. Rekveld, *Ghosts of Luca Pacioli*. Umwelt, observations, 2007.

[18] J. Rekveld, *Light Matters: Ghosts of Luca Pacioli*. Available at http://www.joostrekveld.net/?p=615. May 2017.

[19] J. Sander (ed.), *Albrecht Dürer: His Art in Context*, Prestel, Munich, 2013, pp. 190–191.

[20] C. H. Séquin, *Misinterpretations and Mistakes in Pacioli's Rhombicuboctahedron*. Available at https://people.eecs.berkeley.edu/~sequin/X/Leonardo/pacioli_rco.html. May 2015.

[21] C. H. Séquin and R. Shiau, *Rendering issues in Pacioli's rhombicuboctahedron*. Tech Report EECS-2015–169, June 2015. Available at https://www2.eecs.berkeley.edu/Pubs/TechRpts/2015/EECS-2015–169.html. May 2017.

[22] H. Serras, Mathematics on the "Ritratto di Fra' Luca Pacioli." Available at http://cage.ugent.be/~hs/pacioli/pacioli.html. May 2017.

Who Would Have Won the Fields Medal 150 Years Ago?

Jeremy Gray

Introduction

Hypothetical histories are a way of shedding light both on what happened in the past and on our present ways of thinking. This article supposes that the Fields Medals had begun in 1866 rather than in 1936 and will come to some possibly surprising conclusions about who the first winners would have been. It will, I hope, prompt readers to consider how mathematical priorities change—and how hindsight can alter our view of the mathematical landscape.

Let us imagine that in 1864 the Canadian–American astronomer Simon Newcomb had seen past the horrors of Antietam and Gettysburg to a better world in which mathematics would take its place among other cultural values in the new republic. Let us further suppose that Newcomb, in his optimism for the future, had decided that there should be a prize awarded regularly to young mathematicians who had done exceptional work and that the first awards should be made in August 1866.

What might have motivated Newcomb to thus recognize exemplary mathematics research? Since 1861, Newcomb had worked at the Naval Observatory in Washington, D.C., as an astronomer and professor of mathematics. There Newcomb helped fill the void left when other academics, uncomfortable at being employed by a military institution during a time of war, resigned their positions at the observatory.

The U.S. government tasked Newcomb with determining the positions of celestial objects. It was challenging work for Newcomb, who had had no formal education as a child and had learned mostly from his father before attending the Lawrence Scientific School at Harvard University and studying under Benjamin Pierce. It was this work which gave Newcomb his appreciation of mathematics.

To give the mathematics prize credibility, Newcomb would have realized that he would have to select a panel of sufficiently eminent judges. To find them, he would have to travel to Europe, so let us follow him to Paris in early 1865, the year he turned thirty.

The French were then feeling an unaccustomed inferiority. For two generations, the French had dominated mathematics. Laplace, in the five-volume *Traité de Mécanique Céleste* (1798–1825), had given conclusive reasons to believe that Newtonian gravity rules the solar system and could explain all apparently discordant observations. Lagrange had been the true successor of Euler in many fields, and what little he had set aside Legendre had taken up. These luminaries had been succeeded by Cauchy, Fourier, and Poisson, and Paris had drawn in many of the best young mathematicians from abroad, Abel and Dirichlet among them.

But these great figures had all died by 1865, and news from the German states had the French looking uncomfortably second rate. The influence of Gauss, who had died in 1855, the year before Cauchy, was ever more apparent. Gauss had revivified the theory of numbers—"the higher arithmetic," as he had called it—making it more substantial than Euler or Lagrange had managed to. Closely intertwined with Jacobi's elliptic function theory, the field had been further extended by Dirichlet. Although Jacobi and Dirichlet had both died by 1865, the University of Berlin was flourishing, and, it had to be admitted, the École Polytechnique was not. Apparently the glory days of Napoleon had taught better lessons to those whose countries he had conquered than to the French themselves, who found themselves with a complacent government unable or unwilling to keep up.

Newcomb would make it his business to meet Joseph Liouville, who was the founder and editor of the important *Journal de Mathématiques Pures et Appliquées* and used its pages to keep the French mathematical community aware of what was happening abroad. Liouville had been involved in bringing to France Kummer's discovery of the failure of the prime factorization of cyclotomic integers and his attempts to redefine "prime" to deal with this unexpected setback. Kummer had subsequently won the Grand Prix des Sciences Mathématiques of the Paris Academy of Sciences in 1857 for his work on the subject. As a successful editor, Liouville was the best person to ask about new and exciting mathematicians. And he would be fifty-six in 1865, too old to be considered for the prize himself.

Liouville had done important work on the theory of differential equations (Sturm–Liouville theory), potential theory, elliptic and complex functions, the shape of the Earth, and other subjects. But Newcomb might well have been discouraged on consulting the most recent issues of the *Journal*, because they were full of Liouville's interminable and shallow investigations of the number theory of quadratic forms in several variables. In a sense, what Liouville's *Journal* and the *Journal* of the École Polytechnique showed was that mathematics in France was at a low ebb.

Liouville would surely have praised Charles Hermite highly, had Newcomb prompted him to suggest potential prizewinners. Hermite had studied with Liouville, and together they had come to a number of insights about elliptic functions. It was in this context that Liouville had discovered the theorem that still bears his name: a bounded complex analytic function that is defined everywhere in the plane is a constant. Hermite had gone on to explore the rich world of elliptic functions and in 1858 had used that theory and algebraic invariant theory to show that the general polynomial equation of degree five has solutions expressible in terms of elliptic modular functions. This work had drawn the attention of Kronecker and Brioschi and remained an insight that rewarded further attention. Galois's discovery that the general quintic equation was not solvable by radicals sat in the context of equations that are solvable by precise classes of analytic functions.

But Hermite was forty-two. Newcomb wanted to recognize younger mathematicians to point the way to the future, and he was beginning to realize that he needed more advisors than Liouville alone.

Kummer was the obvious choice. He was less than a year younger than Liouville, and he had worked on a variety of topics before deciding that Gaussian number theory was the rock on which to build a career. He had written on the hypergeometric equation (another Gaussian topic) and as recently as 1863 on a quartic surface with sixteen nodal points. Kummer had succeeded Dirichlet at Berlin in 1855, when Dirichlet moved to Göttingen to succeed Gauss, and had swiftly arranged for Weierstrass to be hired at Berlin. Weierstrass, who had just turned forty then, had only recently burst onto the mathematical scene with his theory of hyperelliptic functions and was now, with Kummer, establishing Berlin as the preeminent place to study mathematics. Furthermore, Kummer, as proof of his administrative ability, had just become dean of the University of Berlin. Newcomb would choose Kummer as the second member of his prize committee.

By now Newcomb was becoming aware that the new state of Italy (unified only in 1861) was also producing important mathematicians. The Italian figures comparable to the German judges Kummer and Liouville were Enrico Betti and Francesco Brioschi, who had both just turned forty. Both had been actively involved in the unification of Italy and now led political lives: Betti had been elected to the Italian parliament in 1862, and Brioschi was to become a senator in 1865. Betti had just become the director of the Scuola Normale Superiore; Brioschi was the founder and director of the Istituto Tecnico Superiore in Milan. Both men were devoted to raising the standard of mathematics in Italy in both schools and universities; both were active in research. Betti, who had become a close friend of Riemann's when he stayed in Italy, was interested in extending Riemann's topological ideas and also worked on mechanics and theoretical physics. Brioschi had done important work in algebra, the theory of determinants, and elliptic and hyperelliptic function theory, and he had taught many of the next generation of Italian mathematicians: Casorati, Cremona, and Beltrami among them. Newcomb, we shall suppose, would have decided that Brioschi was the man to keep him informed of the latest developments in the emerging domain of Italian mathematics.

Should that be enough, or should Newcomb make a trip to Britain? In pure mathematics, this meant a visit to Cayley; in more applied fields, there were several people at Cambridge who might be consulted. The mathematicians Newcomb had already met spoke well of Cayley and respected him as an inventive and well-read mathematician who spoke several languages. He was, along with his friend Sylvester, best appreciated for his exhaustive, and sometimes exhausting, investigations into invariant theory. But it did not seem that there was anyone in England in 1865 who could be considered for the prize, now that Cayley was in his early forties and thus ineligible himself.

Newcomb decided that three judges were already enough: Liouville, Kummer, and Brioschi. It was time to select a prizewinner.

Newcomb already knew one name. Everyone he spoke to told him about Bernard Riemann, a truly remarkable former student of Gauss in Göttingen. Riemann had published a remarkable paper on abelian functions in 1857 that was so innovative that Weierstrass had withdrawn a paper of his own on the subject, saying that he could not proceed until he had understood what Riemann had to put forward. That same year, Riemann had published a very difficult paper on the distribution of the primes, in which he made considerable use of the novel and fundamentally geometric theory of complex analytic functions that he had developed and which was essential to the paper on abelian functions. There was also a paper in real analysis where he had developed a theory of trigonometric series to explore the difficult subject of nondifferentiable functions, and there was talk of a paper in which he was supposed to have completely rewritten the subject of geometry.

Riemann would turn thirty-nine in 1865, so Newcomb could agree that he was still, officially, young. There was the disturbing problem of Riemann's health, however. He suffered from pleurisy and was said to have collapsed in 1862 and to be recuperating in Italy. Newcomb would have to stay informed.

As for the generation born in the 1830s, Liouville, Kummer, and Brioschi might well have given different reports.

Liouville, to his regret, would have had no one to suggest. Kummer, too, would have struggled to name a nominee. His former student Leopold Kronecker had just turned forty, and although there were some promising younger mathematicians at Berlin—Lazarus Fuchs sprang to mind—they had yet to do anything remarkable.

Brioschi, on the other hand, would have been optimistic. He could have suggested the names of Cremona, Casorati, and Beltrami. Cremona was already known for his work on projective and birational geometry, including the study of geometric (birational) transformations, and Brioschi could have assured the panel that Cremona was writing a major paper on the theory of cubic surfaces (it was to share the Steiner Prize in 1866—Kummer was one of the judges—and was published in 1868). Casorati was perhaps the leading complex analyst the Italians had produced, and Beltrami was emerging as a differential geometer in the manner of Riemann.

Over the summer of 1865, Newcomb would have faced a difficult decision. No one disputed Riemann's brilliance, although Kummer reported that Weierstrass was hinting that not all of Riemann's claims were fully established, and Liouville was saying that Hermite was hoping for direct proofs of some results that presently relied on Riemannian methods.

The problem was to decide who else might get the prize. There were several bright young mathematicians, but none of the highest caliber. Should Newcomb announce the existence of the prize, call for nominations, and risk disappointment? Or should he postpone it and give the younger mathematicians a better chance to shine?

And then there was the worrisome matter of Riemann's deteriorating health. Weak and prone to illness, he had spent the summer recuperating near Lake Maggiore and in Genoa and had returned to Göttingen in early October.

Let's suppose that Newcomb decided to postpone the competition for four years.

Riemann died on July 20, 1866, within a month of returning to Lake Maggiore. He was thirty-nine. Among the papers published shortly after his death, the one entitled "The hypotheses that lie at the foundations of geometry" was to inspire Beltrami (born 1835) to publish his "Saggio," in which non-Euclidean geometry was described rigorously in print for the first time. The leading German physicist, Hermann von Helmholtz, was independently converted to the possibilities of spherical geometry and in correspondence with Beltrami came to advocate the possibilities of non-Euclidean ("hyperbolic") geometry as well.

Asked about candidates from Germany, Kummer could now offer three or four. The first was Rudolf Clebsch, who, with his colleague Paul Gordan, had devised an obscure but effective notation for invariants that had led to many new results, and he had applied himself successfully to the study of plane curves. He also had showed that elliptic functions can be used to parameterize cubic curves and in 1864 had begun to extend such ideas to curves of higher genus. Then, in 1868, he had opened the way to extending Riemann's ideas to the study of complex surfaces by defining the (geometric) genus of an algebraic surface.

The second was Lazarus Fuchs, a former student of Kummer's, who was now attached to Weierstrass and seemed poised to extend Riemann's ideas. So too was the third candidate, Hermann Amandus Schwarz, who was using Weierstrass's representations of minimal surfaces to tackle the Plateau problem. He was also beginning to think about the Dirichlet problem.

Then there was the fourth candidate, Richard Dedekind, who was emerging as a number theorist in the tradition of Gauss and Dirichlet. But Kummer, and still more his colleague Kronecker, had their doubts about the highly abstract and not always explicit character of Dedekind's approach. It would seem appropriate to wait.

Liouville, too, would have had a new candidate to put forward as the 1860s came to an end: Camille Jordan. Jordan had published a series of papers that he was now drawing together in his book *Théorie des Substitutions et des Équations Algébriques*. In these papers, and again in the book, he set out a theory of groups of substitutions (permutation groups) of great generality and showed how to use it to derive all of Galois's results systematically. He had then gone on to use it in a number of geometrical settings, finding, for example, the groups of the twenty-seven lines on a cubic surface and Kummer's surface with its sixteen nodal points. He outlined in over three hundred pages a program to find all finite groups. Not everyone was convinced of the need for such a big new idea, but it was bold and rich in applications to topics that were known to be interesting.

And what of James Clerk Maxwell? Could his best work even be called mathematics? He had written on many subjects, but his major 1864 paper, "A dynamical theory of the electromagnetic field," and an 1866 paper in which he suggested that electromagnetic phenomena travel at the speed of light (thus implying that light is just such a phenomenon) displayed a considerable mastery of the difficult mathematics they involved. He had also published his second major paper on the dynamical theory of gases, which did much to establish the statistical approach to physics.

So Newcomb would have had four candidates: Beltrami, Clebsch, Jordan, and Maxwell. Under pressure, Brioschi would have had to admit that Beltrami had published only one remarkable result amid a stream of good ones, mostly in differential geometry. But at least his work was independent of Riemann's, as Cremona would attest. Clebsch was another heir of Riemann's, but he worked in a tradition that was opposed in some ways to Kummer's way of doing things. Jordan was the youngest, and his advocacy of substitution group theory was controversial. Some found it a fine addition to the geometrical way of thinking, and some were to see in it the way to rewrite Galois theory the way Galois might have meant it, but others were to find it needlessly abstract and almost unnecessary even for Galois theory (and others preferred the language of field extensions).

As for Maxwell, electricity and magnetism had been the major topics of mathematical physics for the preceding fifty years, but no one in continental Europe understood Maxwell's ideas. In particular, they found it incomprehensible that electric current was a discontinuity in a field and not the passage of a (possibly mysterious) substance. It was too risky to choose Maxwell.

What might Brioschi, Kummer, Liouville, and Newcomb then decide? From today's perspective, it seems that they should have rewarded (1) conclusive proof of the existence of a new possible geometry of space (when the work of Bolyai and Lobachevskii was almost completely forgotten), and (2) a new and highly abstract structure (the group) that seemed to have many applications, although many important details remained to be published. Beltrami and Jordan might well be our modern choices.

But I suggest that Newcomb and his advisors would have chosen differently. Kummer had a great sympathy for the study of algebraic surfaces and a high opinion of Cremona's work. Brioschi could have agreed that it offered a way into the general study of algebraic geometry, to which Clebsch had also made a contribution, and that the new, non-Euclidean geometry, however remarkable, had yet to lead to new results. If Cremona represented the opening up of a subject that had long challenged mathematicians, then Jordan could embody the spirit of radical innovation, one that also led to insights into geometry. But such a decision would be strongly opposed by Kronecker and Hermite, who were powerful advocates of invariant theory, and their views would be well known to Kummer and Liouville. That would have made Clebsch a contender, not so much for his Riemannian work as for his development of invariant theory.

Newcomb, who was to publish a paper in 1877 on spaces of constant positive curvature, might have pushed hard for Beltrami. But in the 1860s, his own work was firmly in mathematical astronomy, so I suppose that he would have let himself be guided by the counsel of his chosen judges. He surely would have wanted to see that Riemann's ideas were being carried forward, but the citation for Clebsch, the leading advocate of Riemann's ideas, would take care of that. The prizes, I conclude, would have gone to Cremona and Clebsch.

End Note

The real Fields Medal was established in the bequest of John Charles Fields, a Canadian mathematician who had been active in the

International Mathematical Union. It was his wish that the prize be awarded every four years to two young mathematicians, and although he did not define "young," this has come to mean under forty, and that guideline has therefore been preserved here. The first Fields Medals were awarded in 1936 to Lars Ahlfors and Jesse Douglas.

Paradoxes, Contradictions, and the Limits of Science

Noson S. Yanofsky

Science and technology have always amazed us with their powers and ability to transform our world and our lives. However, many results, particularly over the past century or so, have demonstrated that there are limits to the abilities of science. Some of the most celebrated ideas in all of science, such as aspects of quantum mechanics and chaos theory, have implications for informing scientists about what cannot be done. Researchers have discovered boundaries beyond which science cannot go and, in a sense, science has found its limitations. Although these results are found in many different fields and areas of science, mathematics, and logic, they can be grouped and classified into four types of limitations. By closely examining these classifications and the way that these limitations are found, we can learn much about the very structure of science.

Discovering Limitations

The various ways that some of these limitations are discovered is in itself informative. One of the more interesting means of discovering a scientific limitation is through paradoxes. The word *paradox* is used in various ways and has several meanings. For our purposes, a paradox is present when an assumption is made and then, with valid reasoning, a contradiction or falsity is derived. We can write this thus:

$$\text{Assumption} \rightarrow \text{Contradiction}$$

Because contradictions and falsehoods need to be avoided, and because only valid reasoning was used, it must be that the assumption was incorrect. In a sense, a paradox is a proof that the assumption is not a valid part of reason. If it were, in fact, a valid part of reason, then no contradiction or falsehood could have been derived.

A classic example of a paradox is a cute little puzzle called the *barber paradox*. It concerns a small, isolated village with a single barber. The village has the following strict rule: If you cut your own hair, you cannot go to the barber, and if you go to the barber, you cannot cut your own hair. It is one or the other, but not both. Now, pose the simple question: Who cuts the barber's hair? If the barber cuts his own hair, then he is not permitted to go to the barber. But he is the barber! If, on the other hand, he goes to the barber, then he is cutting his own hair. This outcome is a contradiction. We might express this paradox thus:

$$\text{Village with rule} \rightarrow \text{Contradiction}$$

The resolution to the barber paradox is rather simple: The village with this strict rule does not exist. It cannot exist because it would cause a contradiction. There are a lot of ways of getting around the rule: The barber could be bald, or an itinerant barber could come to the village every few months, or the wife of the barber could cut the barber's hair. But all these are violations of the rule. The main point is that the physical universe cannot have a village with this rule. Such playful paradox games may seem superficial, but they are transparent ways of exploring logical contradictions that can exist in the physical world, where disobeying the rules is not an option.

A special type of paradox is called a *self-referential paradox*, which results from something referring to itself. The classic example of a self-referential paradox is the *liar paradox*. Consider the sentence, "This sentence is false." If it is true, then it is false, and if it is false, then because it says it is false, it is true—a clear contradiction. This paradox arises because the sentence refers to itself. Whenever there is a system in which some of its parts can refer to themselves, there will be self-reference. These parts might be able to negate some aspect of themselves, resulting in a contradiction. Mathematics, sets, computers, quantum mechanics, and several other systems possess such self-reference, and hence have associated limitations.

Some of the stranger aspects of quantum mechanics can be seen as coming from self-reference. For example, take the dual nature of light. One can perform experiments in which light acts like a wave, and other experiments in which it acts like a particle. So which is it? The answer is that the nature of light depends upon which experiment is performed. Was a wave experiment performed, or was a particle experiment performed? This duality ushers a whole new dimension into science. In classical science, the subject of an experiment is a closed system that researchers poke and prod in order to determine its properties. Now, with quantum mechanics, the experiment—and more important, the experimenter—become part of the system being measured. By the act of measuring the system, we self-referentially affect it. If we measure for waves, we affect the system so that we cannot measure for particles and vice versa. This outcome is one of the most astonishing aspects of modern science.

The central idea of a paradox is the contradiction that is derived. Where the contradiction occurs tells us a lot about the type of limitation we found. The paradox could concern something concrete and physical. There are no contradictions or falsehoods in the physical universe. If something is true, it cannot be false, and vice versa. The physical universe does not permit contradictions, and hence, if a certain assumption leads to a contradiction in the physical universe, we can conclude that the assumption is incorrect.

Although contradictions and falsehoods cannot occur in the physical universe, they can occur in our mental universe and in our language. Our minds are not perfect machines and are full of contradictions and falsities. We desire contradictory things. We want to eat that second piece of cake and also to be thin. People in relationships simultaneously love and hate their partners. People even willfully believe false notions. Our language, a product of our mind, is also full of contradictions. When we meet a contradiction in mental and linguistic paradoxes, we essentially are able to ignore it because it is not so strange to our already confused minds.

Rational Assumptions

One of the oldest stories about a limitation of science in classical times concerns the square root of two, $\sqrt{2}$. In ancient Greece, Pythagoras and his school of thought believed that all numbers are whole numbers or ratios of two whole numbers, called *rational numbers.* A student of Pythagoras, Hippasus, showed that this view of numbers is somewhat limited and that there are other types of numbers. He showed that $\sqrt{2}$ is not a rational number and is, in fact, an *irrational number* (generally defined as any number that cannot be written as a ratio or fraction).

We do not know how Hippasus showed that the square root of two is irrational, but there is a pretty and simple geometric proof (attributed to American mathematician Stanley Tennenbaum in the 1950s) that is worth pondering. The method of proof is called a *proof by contradiction,* which is like a paradox. We are going to assume that $\sqrt{2}$ is a rational number and then derive a contradiction:

$$\sqrt{2} \text{ is a rational number} \rightarrow \text{Contradiction}$$

From this contradiction, we can conclude that $\sqrt{2}$ is not a rational number.

First, assume that there are two positive whole numbers such that their ratio is the square root of two. Let us assume that the two smallest such whole numbers are a and b. That is, $\sqrt{2} = a/b$.

Squaring both sides of this equation gives us $2 = a^2/b^2$. Multiplying both sides by b^2 gives us $2b^2 = a^2$.

From a geometric point of view, this equation means that there are two smaller squares whose sides are b, and they are exactly the same size as a large square whose sides are a. That is, if we put the two smaller squares into the larger square, they will cover the same area.

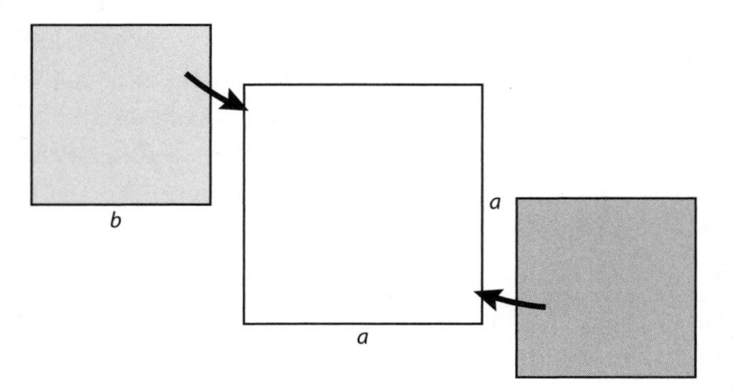

But when we actually place the two smaller squares into the larger, we find two problems. Firstly, we are missing two corners.

Secondly, there is overlap in the middle. So for the area of the larger square to equal the areas of the two smaller squares, the missing areas must equal the overlap. That is, 2(missing) = overlap.

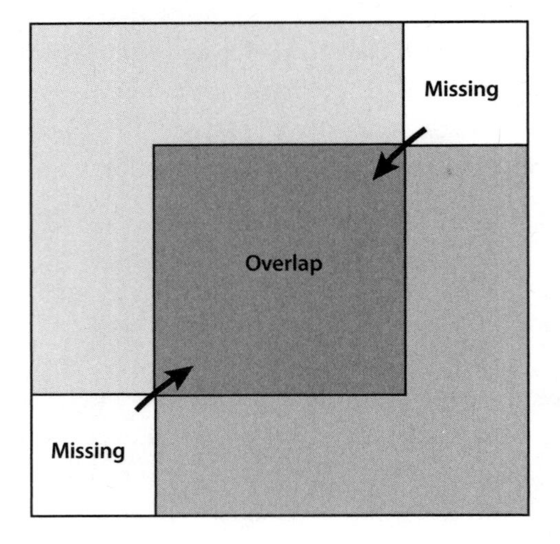

But wait. We assumed that *a* and *b* were the smallest such numbers with which this result can happen; now we find smaller ones. So this result is a contradiction. There must be something wrong with our assumption that *a* and *b* are whole numbers. And thus the square root of two is not a rational number, but is irrational. Hippasus had shown that there was a number that did not follow the dictates of Pythagoras's science.

The followers of Pythagoras were fearful that the conclusion of Hippasus would be revealed and people would see the failings of the Pythagoras philosophy and religion. Legend has it that the other students of Pythagoras took Hippasus out to sea and threw him and his irrational ideas overboard.

We cannot always be so cavalier about ignoring contradictions and falsities in human thought and speech. There are times when we must be more careful. Science is a human language that measures, describes, and predicts the physical world. Because science is constructed to mimic the contradiction-free physical universe, it also must not contain contradictions. Similarly, in mathematics, which is formulated by looking at the physical world, we cannot derive any contradictions. If we did, it would not be mathematics. When a paradox is derived in science or mathematics, it cannot be ignored, and science and mathematics must reject the assumption of the paradox. As an example of such a paradox, if we assume that the square root of two is a rational number, we get a contradiction (*see* box called Rational Assumptions). In this case, we must not ignore the paradox but rather proclaim that the square root of two is not a rational number.

In addition to paradoxes, there are other ways of discovering limitations. Simply stated, one can piggyback off of a given limitation that shows that a certain phenomenon cannot occur, to show that another, even harder phenomenon also cannot occur. A simple example: When you are out of shape and climb four flights of steps, you will huff and puff. We can write this activity and its result as

Climb four flights → Huff and puff

It is also obvious that if someone climbs five flights of steps, they also have climbed four flights of steps, that is,

Climb five flights → Climb four flights

Combining these two implications gives us

Climb five flights → Climb four flights → Huff and puff

We conclude with the obvious observation that if you huff and puff after climbing four flights of steps, you will definitely huff and puff after climbing five flights of steps.

To generalize this simple example, assume that a limitation is found through a paradox:

Assumption-A → Contradiction

Thus, Assumption-A is impossible. If we further show that

Assumption-B → Assumption-A,

we can combine these two implications to get

Assumption-B → Assumption-A → Contradiction

This result shows us that because Assumption-A is impossible, then Assumption-B, is also impossible.

With these methods of finding various limitations, we can categorize the four actual classes of limitations.

Physical Limitations

The first and most obvious type of limitation is one that says certain physical objects or processes cannot exist, like the village in the barber paradox.

The Shortest Route

The Traveling Salesperson Problem is an easily stated computer problem that is an example of a practical limitation. Consider a traveling salesperson who wants to find the shortest route, from all possible routes, that will visit 10 different specified cities. There are many different possible routes the salesperson can take. There are 10 choices for the first city, nine choices for the second city, eight choices for the third city, and so on, down to two choices for the ninth city, and one choice for the tenth city. In other words, there are $10 \times 9 \times 8 \times \ldots \times 2 \times 1 = 10! = 3{,}628{,}800$ possible routes. A computer (in some cases) would have to check all these possible routes to find the shortest one. Using a modern computer, the calculation can be done in a couple of seconds. But what about going to 100 different cities? In some cases, a computer would have to check $100 \times 99 \times 98 \times \ldots \times 2 \times 1 = 100!$ possible routes, which results in a 157-digit-long number: 93,326,215,443,944,152,681,699,238,856,266,700,490,715,968,264,381,621,468,592,963,895,217,599,993,229,915,608,941,463,976,156,518,286,253,697,920,827,223,758,251,185,210,916,864,000,000,000,000,000,000,000,000 potential routes.

For each of these potential routes, the computer would have to calculate how long the route takes, and then compared all of them to find the shortest route. A modern computer can check about a million routes in a second. That computation works out to take 2.9×10^{142} centuries, a long time to find the solution.

Such a problem will not go away as computers get faster and faster. A computer 10,000 times faster, able to check 10 billion possible routes in a second. This will still take 2.9×10^{138} centuries. Similarly, having many computers working on the problem will not help too much. Physicists tell us that there are 10^{80} particles in the visible universe. If every one of those particles were a computer working on our problem, it would still take 10^{62} centuries to solve it. The only thing that possibly can help this problem is finding a new algorithm to figure out the shortest route without looking through all the possibilities. Alas, researchers have been looking for decades for such a magic algorithm. They have not found one, and most computer scientists believe that no such algorithm exists.

The Traveling Salesperson Problem can be solved for small inputs or for certain types of inputs. Even for large inputs, a program can be written that will solve it, but the program will demand an unreasonable amount of time to determine the solution. Although there does exist a shortest possible route, the knowledge of that route is inherently beyond our ability to ever know, making it a practical limitation.

Another example of a physical process that is impossible is time travel into the past. This limitation is usually shown through a self-referential paradox that is called the *grandfather paradox*. In it, a person goes back in time and kills his bachelor grandfather. Thus the father and the time traveler himself will not be born—and hence the time traveler will not be able to kill his grandfather. One need not be homicidal to obtain such a paradox: In the 1985 movie *Back to the Future*, the main character starts to fade out of existence because he traveled back in time and accidentally stopped his mother and father from getting married. A time traveler need only go back several minutes and restrain the earlier version of himself from getting into the time machine.

What is different about events in time travel that cause these paradoxes? Usually, an event affects another, later event: If I eat a lot of cake, I will gain weight. With the time travel paradox, an event affects itself. By killing his bachelor grandfather, the time traveler ensures that he cannot kill his bachelor grandfather. The event negates itself. The simple resolution to the grandfather paradox is that, in order to avoid contradictions, time travel is impossible. Alternatively, if perchance time travel is possible, it is impossible to cause such a contradiction.

Another example of a limitation that shows the impossibility of a physical process is the *halting problem*. Before engineers actually built modern computers, Alan Turing showed that there are limitations to what computers can perform. In the 1930s, before helping the Allies win World War II by breaking the Germans' Enigma cryptographic code, Turing showed what computers cannot do by way of a self-referential paradox. As anyone who deals with computers knows, sometimes a computer "gets stuck" or goes into an "infinite loop." It would be nice if there were a computer that could determine whether a computer will get stuck in an infinite loop. Turing showed that no such computer could possibly exist. He showed that if such a computer could exist, he would make a computer that would negate its own "haltingness." Such a program would perform the following task: "When asked if I will halt or go into an infinite loop, I will give the wrong answer." However, computers cannot give wrong answers because they do exactly what their instructions tell them to do; hence we have a contradiction, which occurs because of the assumption that we made about a computer that can determine whether any computer will go into an infinite loop. That assumption is incorrect. Many other problems in computer science, mathematics, and physics are shown to be unsolvable by piggybacking off the fact that the halting problem is unsolvable.

There are many other examples of physical limitations. For instance, Einstein's special theory of relativity tells us that a physical object cannot travel faster than the speed of light. And quantum theory tells us that the action of individual subatomic particles is probabilistic, so no physical process can predict how a given subatomic particle will act.

Mental Construct Limitations

Recall that although our minds are full of contradictions, we must, when dealing with science and mathematics, steer clear of them, and that means restricting certain mental and linguistic activities.

In the first years of elementary school, we learn an easy mental construct limitation: We are not permitted to divide by zero. Despite the reasons for this rule being so obvious to us now, let us justify it.

Consider the equation $3 \times 0 = 4 \times 0$. Both sides of the equation are equal to zero and hence the statement is true. If you were permitted to divide by zero, you could cancel out the zeros on both sides of the equation and get $3 = 4$. This outcome is a clear falsehood that must be avoided.

A more advanced result in which one sees the mental construct limitation more clearly is in *Russell's paradox*. In the first few years of the 20th century, British mathematician Bertrand Russell described a paradox that shook mathematics to its core. At the time, it was believed that all of mathematics could be stated in the language of *sets,* which are collections of abstract ideas or objects. Sets can also contain sets, or even have themselves as an element. This idea is not so far-fetched: Consider the set of ideas that are contained in this article. That set contains itself. The set of all sets that have more than three elements contains itself. The set of all things that are not red contains itself. The fact that sets can contain themselves makes the whole subject ripe for a self-referential paradox.

Russell said that we should consider all sets that do not contain themselves and call that collection R (for Russell). Now simply pose the question: Does R contain R? If R does contain R, then as a member of R that is defined as containing only those sets that do not contain themselves, R does not contain R. On the other hand, if R does not contain itself, then, by definition, it belongs in R. Again we arrive at a contradiction. The best method of resolving Russell's paradox is to simply declare that the set R does not exist.

What is wrong with the collection of elements we called R? We gave a seemingly exact statement of which types of objects it contains: "those sets that do not contain themselves." And yet, we have declared that this collection is not a legitimate set and cannot be used in a mathematical discussion. Mathematicians are permitted to discuss the green apples in my refrigerator but are not permitted to discuss the collection R. Why? Because the collection R will cause us to arrive at a contradiction. Mathematicians must restrain themselves because we do not want contradictions in our mathematics.

In 1931, Austrian mathematician Kurt Gödel, then 25 years old, proved one of the most celebrated theorems of twentieth-century mathematics. Gödel's Incompleteness Theorem shows that there are statements in mathematics that are true but are not provable. Gödel showed this result by demonstrating that mathematics can also talk about itself. Mathematical statements about numbers can be converted into numbers. Using this ability to self-reference, he formulated a mathematical statement that essentially says: "This mathematical statement is not provable." It's a mathematical statement that negates its own provability. If you analyze this statement carefully, you realize that it cannot be false

Estimating the Unsolvable

How much is beyond our ability to solve? In general, such things are hard to measure. However, in computer science there is an interesting result along these lines. We all know of many different tasks that computers perform with ease. However, there are many problems that are beyond the ability of computers. We can examine whether there are more solvable problems than unsolvable problems.

First, a bit about infinite sets. Mathematicians have shown that there are different levels of infinity. The smallest infinity corresponds to the natural numbers: {0, 1, 2, 3, ...}. We say that this set of numbers is "countably infinite." Although we can never finish counting the natural numbers, we can at least begin listing them. In contrast, the set of all real numbers—that is, numbers such as −473.4562372... and pi—are "uncountably infinite." We cannot even begin to count them. After all, what is the first real number after 0? 0.000001? What about 0.0000000001? It can be shown easily that uncountably infinite sets are vastly larger than countably infinite sets.

Now let us turn to computers that solve problems. There are a countably infinite number of potential computer programs for solvable computer problems. In contrast, there are uncountably infinite computer problems. If one takes all the uncountably infinite computer problems and subtracts the countably infinite solvable problems, one is left with uncountably infinite unsolvable problems. Thus, the overwhelmingly vast majority of computer problems cannot be solved by any computer. Computers can only solve a small fraction of all the problems there are.

(because then it would be provable), and hence it must be true and contradictory. But since it is true, it must also be unprovable. Gödel showed that not everything that is true has a mathematical proof.

Throughout mathematics and science, there are many other examples of mental construct limitations. For instance, one cannot consider the square root of 2 to be a rational number (*see* box called Rational Assumptions). Zeno's famous paradoxes involving such conundrums as motion being an illusion, can also be seen as examples of mental construct limitations.

Practical Limitations

So far, we have seen limitations that show it is impossible for something or some process (physical or mental) to exist. In a practical limitation, we are dealing with things that are possible, albeit extremely improbable. That is, it is impossible to make some prediction or find some solution in a reasonable amount of time or with a reasonable amount of resources.

The classical example is the butterfly effect from chaos theory. The phrase comes from the title of a 1972 presentation by mathematician Edward Lorenz of the Massachusetts Institute of Technology: "Predictability: Does the flap of a butterfly's wings in Brazil set off a tornado in Texas?" Lorenz was a meteorologist and a mathematician. He was discussing the fact that weather patterns are extremely sensitive to slight changes in the environment. A small flap of a butterfly's wing in Brazil might cause a change that causes a change that eventually causes a tornado in Texas. Of course, one should not go out and kill all the butterflies in Brazil; the butterfly flap might instead send a coming tornado off course and save a Texas city. The point of the study is that because there is no way we can keep track of the many millions of butterflies in Brazil, we can never predict the paths of tornados or of the weather in general. This thought experiment shows a limitation of our predictive ability.

Many other problems from chaos theory show limitations. Predicting tomorrow's lottery numbers is also beyond our ability. If you wanted to know the numbers, you would have to keep track of all the atoms in the bouncing ball machine—far too many for us to ever be able to do.

Perpetual motion machines are another example of a practical limitation. There is essentially no way that one can make a machine that will continue to move without losing all its energy. One might be tempted to say that this limitation is really a physical one because it says that a perpetual motion machine cannot exist in the physical universe. But by the second law of thermodynamics, it is extremely improbable for there to be a machine that does not dissipate its energy. Improbable, but not impossible.

The theory of thermodynamics and statistical mechanics is about large groups of atoms and the heat and energy they can create. Because in such systems there are too many elements to keep track of, the laws in such theories are given as probabilities, and are ripe for finding other examples of practical limitations. In computer science, an example of a problem that is theoretically solvable, but for large inputs will never practically be solved, is called the *traveling salesperson problem* (*see* box called The Shortest Route). There are many more.

Limitations of Intuition

The fourth type of limitation is more of a problem with the way we look at the world. Science has shown that our naive intuition about the universe that we live in needs to be adjusted. There are many aspects of reality that seem obvious, but are, in fact, simply false.

One of the most shocking examples of this false perception comes from Einstein's special theory of relativity. The notion of *space contraction* says that if you are not moving and you observe an object moving near the speed of light, then you will see the object shrink. This observation is not an optical illusion: The object actually shrinks. Similarly, the phenomenon of *time dilation* says that when an object moves close to the speed of light, all the processes of the object will slow down. Of course, an observer traveling with the object will see neither space contraction nor time dilation. Thus, our naive view that objects have fixed sizes and processes have fixed duration is faulty.

Some of the most counterintuitive aspects of modern science occur within quantum mechanics. Since the beginning of the past century, physicists have been showing that the subatomic world is an extremely strange place. In addition to finding that the properties of things (such as a photon acting like a wave or a particle) depend on how they are measured, researchers have found that rather than a particle having a single position, it can be in many places at one time, a property called *superposition*. Indeed, not only position, but many other properties of a subatomic particle, might have many different values at the same time. Heisenberg's uncertainty principle tells us that objects do not have definitive properties until they are measured. A famous concept called Bell's theorem shows us that an action here can affect objects across the universe, which is called *entanglement*.

One might think that mathematics is always intuitive and that our intuitions in that field at least might never need to be adjusted. But this assumption is also not true. In the late nineteenth century, German mathematician Georg Cantor, a pioneer in set theory, showed us that our intuition about infinity is somewhat troublesome. The naive view is that all the infinite sets are the same size. Cantor showed that in fact there are many different sizes of infinite sets. (*See* box called Estimating the Unsolvable.)

In the sciences, whenever there is a paradigm shift, all of our ideas about a certain subject have to be readjusted. We have to look at phenomena from a new viewpoint.

The Unknowable

The classification of the limitations of science is only beginning, and many questions still arise. Is this classification complete, or are there other types of limitations? Is there a subclassification of each of the classes? How do the methods of finding the limitations correspond to the types of limitation? Are there results that are in more than one classification? Because some of the results in the other classes might also be counterintuitive, there might be some overlap between categories.

How widespread is this inability to know? Most scientists work in areas in which progress in knowing happens every day. What about what cannot be known? In general, the concept is hard to measure. There are reasons to believe that there is a lot more "out there" that we cannot know than what we can know. (*See* box called "Estimating the Unsolvable" for such a calculation in computer science.) Nevertheless, it is hard to speculate. Isaac Newton said, "What we know is a drop, what we don't know is an ocean." Similarly, Princeton University theoretical physicist John Archibald Wheeler is quoted as saying, "As the island of knowledge grows, so does the shore of our ignorance." Newton and

Wheeler were talking about what we do not know. What about what we cannot know?

Most of the limitations discussed here are less than a century old, a very short time in the history of science. As science progresses, it will become more aware of its own boundaries and limitations. By looking at these limitations from a unified point of view, we will be able to compare, contrast, and learn about these many different phenomena. We can understand more about the very nature of science, mathematics, computers, and reason.

Bibliography

Barrow, J. D. 1999. *Impossibility: The Limits of Science and the Science of Limits.* Oxford, U.K.: Oxford University Press.

Chaitin, G. 2002. "Computers, paradoxes and the foundations of mathematics." *American Scientist* 90:164–171.

Cook. W. J. 2012. *In Pursuit of the Traveling Salesman: Mathematics and the Limits of Computation.* Princeton, NJ: Princeton University Press.

Dewdney, A. K. 2004. *Beyond Reason: 8 Great Problems That Reveal the Limits of Science.* Hoboken, NJ: Wiley.

Tavel, M. 2002. *Contemporary Physics and the Limits of Knowledge.* New Brunswick, NJ: Rutgers University Press.

Yanofsky, N. S. 2013. *The Outer Limits of Reason: What Science, Mathematics, and Logic Cannot Tell Us.* Cambridge, MA: MIT Press.

Stairway to Heaven: The Abstract Method and Levels of Abstraction in Mathematics

Jean-Pierre Marquis

One of the distinctive features of twentieth-century mathematics is the rise and systematic use of the abstract method. This method changed dramatically both the very object of mathematics and its methods. Mathematics suddenly referred to groups, rings, fields, metric spaces, topological spaces, vector spaces, Banach spaces, manifolds, lattices, and categories, and it relied on the abstract method for its presentation and development. The abstract method is at the core of what came to be known as "modern mathematics." As a by-product, the abstract method opened the door to the idea that there are levels of abstraction in mathematics: there are parts of mathematics that are more abstract than others. Mathematicians themselves speak that way spontaneously. Here are three representative passages found in textbooks:

> However, as we began to think about the task at hand . . ., we decided to organize the material in a manner quite different from that of our earlier books: a separation into two levels of abstraction . . . [12, p. xii]

> This text attempts a different approach, letting the abstract concepts emerge gradually from less abstract problems about geometry, polynomials, numbers, etc. This is how the subject evolved historically. This is how all good mathematics evolves—abstraction and generalization is forced upon us as we attempt to understand the "concrete" and the particular [27, p. 3].

> This book is intended as an introduction to that part of mathematics that today goes by the name of abstract algebra. The term "abstract" is a highly subjective one; what is abstract to one person is very often concrete and down-to-earth to another, and vice versa. In relation to the current research activity in algebra, it could be described as "not too abstract"; from the point of view of someone schooled in the calculus and who is seeing the present material for the first time, it may very well be described as "quite abstract" [11, pp. 1–2].

The quotes state without any further explanation that there are levels of abstraction. The second quote goes somewhat further and asserts that the levels of abstraction arose historically and that abstraction was "forced" on mathematicians. It seems that for these quoted mathematicians, the proposal that there are ideas, theories, and concepts that are more abstract than others is obvious and does not need any further clarification or discussion. It is as if the claim is based on a common, clear, and widespread mathematical experience.

The last quote has a different flavor. The author suggests that "being abstract" is a subjective term. As I hope to show in this paper, I think that this claim is plain wrong. I submit that the author is confusing familiarity with a subject with its abstract character.

My goal in this paper is to clarify what mathematicians might mean by levels of abstraction. I believe that the claim can be taken literally and that it is not a subjective matter. It is, as I will argue, a consequence of the abstract method itself. The paper will therefore proceed as follows. In the next section, I will provide an overview of what I take to be the main components and properties of the abstract method in mathematics. I will then propose a definition of levels of abstraction based on some of the elements of that method. Finally, in the last section, I will explore some of the philosophical consequences of this approach.[1]

The Abstract Method: Basic Moments

I have elsewhere presented some of the historical roots and moments of the abstract method in mathematics [20]. It is nonetheless necessary to recapitulate and develop from a different perspective the conceptual ingredients of my analysis.

Informally, the abstract method can be described as follows. It is made up of three different moments:[2]

1. The first moment is the identification of invariant properties in theories, which seem to be essentially incompatible at first sight.
2. The second moment is the systematic ignorance of some of the specific properties of the objects of each theory (this is the step that justifies the name "abstraction" since one has to "subtract" key properties of the objects of the given theories) and the presentation in an appropriate language of the invariant properties.
3. The last moment is the identification of a criterion of identity on the basis of the properties chosen and, thus, the introduction of a new type of entities, the abstract entities, and the exploration of the properties of these new entities.

Let me reformulate these moments, using a somewhat more formal set-up.[3] The abstract method comprises three basic parts that are related systematically and thus make a whole. The first part is what I call a *domain of significant variation*. It is made up of at least three essentially different theories, denoted by $\mathbf{T}_1$, $\mathbf{T}_2$, and $\mathbf{T}_3$. The second part is a *method of presentation* together with a *method of development*, which was historically identified with the axiomatic method and which yields a *new* theory, denoted by $\mathbf{T}_{Abs}$. The third part is a *new criterion of identity* for the emerging entities, $X \simeq_{\mathbf{T}_{Abs}} Y$.

The links between the parts are as follows. Between the domain of significant variation and the method of presentation and development, we have what I call the *formalist stance*, and it consists of identifying the invariant properties within the domain of variation and forgetting what the symbols used actually refer to. Between the method of presentation and the criterion of identity, we have the *extraction of a new criterion of identity* together with the understanding of how it applies, that is, using the new criterion in a way that the following form of Leibniz's principle applies:

$$\text{If } P(X) \text{ and } X \cong_{\mathbf{T}_{Abs}} Y, \text{ then } P(Y).$$

Here is a diagram capturing the components and their relations.

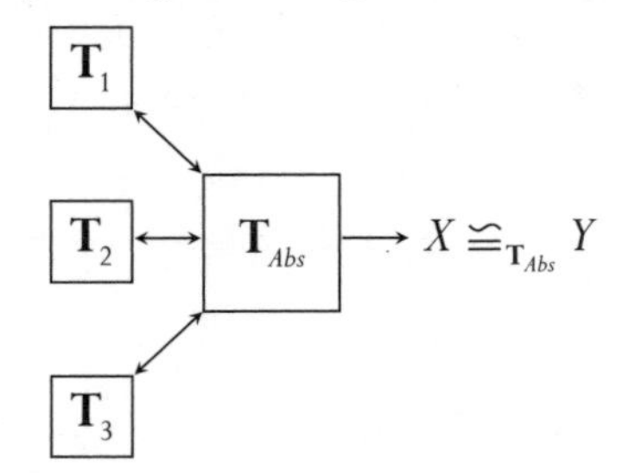

One simple historical example might be useful before I explain the diagram, and thus the method, in more detail. One of the first cases of an extraordinarily successful abstraction in the history of modern mathematics is certainly that of metric spaces, introduced by Fréchet around 1906 in the context of functional analysis.[4] At the turn of the century, mathematicians thought of manifolds as subspaces of spaces of real or complex points. Fréchet, for instance, was dealing with the usual manifolds, namely $\mathbb{R}, \mathbb{R}^2, \ldots, \mathbb{R}^n, \ldots, \mathbb{C}, \mathbb{C}^2, \ldots, \mathbb{C}^n$ together with functions between them on the one hand, and infinite-dimensional functional spaces together with operators between them on the other hand. In his thesis, Fréchet gives the following four examples of functional spaces (see [8] or [28]).

1. Let J be a closed interval of the real line $\mathbb{R}$ and consider the space $\mathbb{R}^J$ of continuous functions $f: J \to \mathbb{R}$. A metric on R^J is defined by
$$d(f, g) = max(|f(x) - g(x)|) \qquad \forall x \in J.$$
2. Consider the space $E_\infty = \mathbb{R}^{\mathrm{N}}$ of infinite sequences $x = (x_1, x_2, \ldots)$ of real numbers. A metric on E_∞ is given by
$$d(x,y) = \sum_{n=1}^{\infty} \frac{1}{n!} \frac{|x_n - y_n|}{1 + |x_n - y_n|}.$$
3. A space of parametrized curves in $\mathbb{R}^3$ with the standard Euclidean metric between points. Using the latter, Fréchet defines a metric between the curves.
4. Finally, let A be a complex plane region whose boundary consists of one or more contours. Let $\{A_n\}$ be a sequence of bounded regions such that $A_n \subset intA_{(n+1)}$ and $A_n \subset int(A)$ and

such that any given bounded region in the interior of A is in the interior of some A_n for n sufficiently large. Consider the space $\{f: int(A) \rightarrow \mathbb{C} \mid f \text{ is holomorphic}\}$, and let

$$M_n(f,g) = max|f(z) - g(z)|$$

when z is in the closure of A_n.

The metric between two such functions is then defined by

$$d(f,g) = \sum_{n=1}^{\infty} \frac{1}{n!} \frac{M_n(f,g)}{1 + M_n(f,g)}.$$

Although I haven't described the third example in detail, I hope that it is nonetheless clear that these examples are radically different from one another and, perhaps even more so, from the spaces of points $\mathbb{R}^n$ and $\mathbb{C}^n$. I submit that if we did not know about the metric involved in each case, we might not see the invariant features involved. Indeed, we are accustomed to attribute certain properties to real functions, e.g., continuity, differentiability, roots, maximum, minimum, and we represent the graph of a function as a one-dimensional path in the codomain, thus as something that necessarily has a length. We think of a real function as a systematic relation of dependence between two or more properties, as a quantity that varies according to a certain pattern or whose variation depends on another variation. A function is essentially thought of as dynamic. The four examples given by Fréchet are of this kind. A (real) point is, well, a point. It has none of the properties of a function. Thus, the properties of the elements of R and even R^n are incommensurable with the properties of the elements of a function space. I want to insist on the fact that given the properties of functions and given that we think of functions with their properties, it is hard to conceive of a *space* of functions, that is, treating the latter as being points. It is as if we were trying to think of the properties of functions and forget about them at the same time. Of course, as soon as we have succeeded in thinking of them as spaces, we stumble on what is certainly seen as being the main difference between these spaces and the usual spaces of points: the examples given here are *infinite* dimensional. Thus, we also have two different types: finite dimensional spaces on one side and infinite dimensional spaces on the other.

I want to emphasize the fact that to see the invariant features between the different theories, one has to *forget* or *ignore* essential aspects of the objects and their properties involved. One has to ignore key properties of functions, of series, of the complex numbers, etc. One of the ways to succeed in this operation is to concentrate on the formalism, the symbols, and the operations on these symbols. This posture is what I call the *formalist stance*. Historically, it is easy to find mathematicians emphasizing the necessity of this stance. Here is one clear illustration:

> In the following an attempt is made to present Galois theory of algebraic equations in a way which includes equally well all cases in which this theory might be used. Thus we present it here as a direct consequence of the group concept illuminated by the field concept, as a *formal structure completely without reference* to any numerical interpretation of the elements used. (Weber 1893, p. 521, quoted by [4, p. 36])[my emphasis].

Weber says clearly that in this case, one has to forget what the symbols stand for and concentrate on the formal structure, on the relations and operations that exist between the objects. The same posture applies to the case of metric spaces: one has to forget the idiosyncrasies of the elements of the various spaces, what the symbols refer to, and focus on the invariant structural features that these domains share. Notice that if one does not move to the next step, namely, the extraction of an appropriate criterion of identity for new objects, then the only coherent philosophical position left is a variant of formalism.

Thus, the next step consists of finding the proper language and the right properties to present the invariant features involved in the domain of significant variation. Historically, the axiomatic method seemed perfect for the purpose. So much so that mathematicians often *identify* the axiomatic method with the abstract method. I will come back to this point later.

After the invariant properties have been identified, it is only natural to investigate the logical consequences of these properties and verify that it is possible to develop relevant and useful mathematics from them. Then, and only then, a criterion of identity for the new, abstract entities can be extracted. In other words, the criterion of identity cannot be given *a priori*, but is derived from the theory.[5]

Let us now look at these components and their relations one by one.

Domain of Significant Variation

> Thus, owing to its abstract foundation, modern algebra winds a unifying band of method *around essentially different things* and in this way contributes to the required organic and systemic unification of mathematics [10, pp. 18–19][my emphasis].

In the foregoing diagram, I have included *three, essentially different* theories. I claim that these two elements, that is, the number of theories and the fact that they differ in an essential manner, are necessary and sufficient for the abstract method to be used in a particular case.

The number three is not accidental. It seems to play a key role in the method itself. I have to emphasize that I present this claim as an empirical hypothesis. Although with hindsight, there are good reasons to believe that the number of theories is indeed necessary, it came to me *a posteriori*, by examining the historical record. Let me start with a long quote from a historian of mathematics who happened to come to the same conclusion for a specific important case.

> It is obviously true that the concept of a permutation group derived from the development of the theory of algebraic equations and from Galois theory. But this development, associated with the names of A. Vandermonde, J.-L. Lagrange, P. Ruffini, N. H. Abel, E. Galois, J.-A. Serret, and C. Jordan, is just *one* of the historical roots of group theory. The mathematical literature of the nineteenth century, and especially the work of decisive importance for the evolution of the abstract group concept written at the century's end, make it abundantly clear that that development had *three* equally important historical roots, namely, the theory of algebraic equations, number theory, and geometry. Abstract group theory was the result of a gradual process of abstraction from implicit and explicit group-theoretic methods and concepts involving the interaction of its three historical roots. I stress that my inclusion of number theory and geometry among the sources of causal tendencies for the development of abstract group theory

> is grounded in the historical record and is not the result of a backward projection of modern group-theoretic thought ([31, p. 16]).

This is but one case. What is striking is that when one looks at the development of ring theory and field theory, one finds a similar situation. (See, for the case of ring theory, [14, 5] and, for the case of field theory, [15].) As we have seen, Fréchet took the time to present four different function spaces, as well as the usual manifolds. What is even more striking is that the abstract method, conceived as a genuine method in algebra, arose only after there were three different cases where the method turned out to be fruitful: groups, rings, and fields.

We can also find *theoretical* reasons that make the claim *a priori* plausible. Indeed, when there are two theories, it seems reasonable to think that one will consider either a simple generalization or an analogy between them. When there are at least three essentially different theories, then the essential differences make it difficult to believe that a simple generalization is possible, and the fact that there are at least three brings us immediately outside the realm of analogies.

The idea of a *significant* variation can be expressed more formally as follows. Each theory $\mathbf{T}_i$ has its own criterion of identity $a \simeq_{\mathbf{T}i} b$. Therefore, we are dealing with (at least) three distinct criteria of identity satisfying Leibniz's principle[6]:

1. In $\mathbf{T}_1$, if $P(\mathbf{a})$ and $\mathbf{a} \simeq_{\mathrm{T1}} \mathbf{b}$, then $P(\mathbf{b})$.
2. In $\mathbf{T}_2$, if $Q(\mathfrak{f})$ and $\mathfrak{f} \simeq_{\mathrm{T2}} \mathfrak{g}$, then $Q(\mathfrak{g})$.
3. In $\mathbf{T}_3$, if $R(\alpha)$ and $\alpha \simeq_{\mathrm{T3}} \beta$, then $R(\beta)$.

These are incompatible in the sense that one would not, for instance, necessarily attribute certain properties, say $P_1, \ldots, P_n$ to the objects α_i of $\mathbf{T}_3$ and conversely, one would not apply the properties $R_1, \ldots, R_k$ to the objects $\mathbf{a}_i$ of $\mathbf{T}_1$. In fact, in some cases, even if one would try to apply, let us say P_i to α_j, one would in some cases consider $\neg P_i$ instead. Thus, to abstract properly, it is necessary to be able to ignore, forget, or subtract some of the properties of the objects in the given theories. Furthermore, the properties that have to be subtracted are not necessarily the same in each case. I submit that this is a real cognitive stumbling block to abstraction in mathematics. A clear empirical hypothesis could be formulated and be tested on students learning vector spaces, group theory, topology, or any other similar theory.

In other words, the epistemic attention has to shift from certain pregnant features of the objects under study to the invariant elements involved across the various theories. The latter are usually operations or relations between these objects. Again, I want to emphasize that at this stage, one could stop and either work in a purely formal fashion or consider that one is doing algebra in the classical sense of that word, that is, working on generalized arithmetic operations. As we have indicated elsewhere (see [20]), it is easy to find this attitude in the nineteenth century, especially when one looks at the development of algebra during that period. The situation changes radically after it is clear that it is possible to consider a new type of entities supporting these properties and relations. After the invariant features have been identified, recognized as such, and confirmed as being the same in all cases, then a criterion of identity can be formulated. The invariant component is from then on circumscribed clearly and independently of the original entities. These are seen to be instances of these new types and are studied as such, that is, there is a shift of attention from the old criterion of identity and its associated properties to the new criterion and its associated properties.

Very often, it is then possible to discover and construct new, unforeseen instances of these new abstract entities. Thus, the domain of variation can expand and is never fixed once and for all. In more philosophical terms, after the new types have been fixed, known examples become tokens of the type, and new, unforeseen tokens can be constructed or discovered.

The Axiomatic Method and the Abstract Method

From a historical perspective, the abstract character of contemporary mathematics can be attributed, at least in part, to the axiomatic method. However, many mathematicians *identify* the abstract method with the axiomatic method. To wit:

> The *abstract or postulational* development of these systems must then be supplemented by an investigation of their "structure." ([17, p. 18], quoted also by [4, p. 258]) [my emphasis].

Mac Lane's formulation is typical of the period: the abstract and the axiomatic seem to be interchangeable. This attitude still prevails today:

> It is abstraction—more than anything else—that characterizes the mathematics of the twentieth century. There is both power and elegance in the axiomatic method, attributes that can and should be appreciated by students early in their mathematical careers and even if they happen to be confronting contemporary abstract mathematics in a serious way for the very first time ([30, p. ix]).

Again, the author writes as if the axiomatic method and the abstract method were the same thing. The axiomatic method is but one method of presentation and development of the invariant content of theories in the domain of variation. The axiomatic method played a key role in the construction of the abstract method, but the latter is not the same thing as the axiomatic method.

It is extremely easy to convince oneself that the axiomatic method has nothing to do, by itself, with the abstract. Indeed, one can use the axiomatic method to present a theory about a unique set of entities, as it was probably conceived to present Euclidean geometry. One can even use the axiomatic method to present a physical theory, e.g., a theory about the concrete world. Thus, as such, it is hard to see how one can identify the abstract with the axiomatic.

The fact is that the axiomatic method has a long history and therefore has acquired various, even incompatible, roles in mathematics and in foundational studies. Let us briefly recapitulate some of the functions of the axiomatic method.

Let us start with Hilbert:

> Hilbert's own use of the axiomatic method involved, by definition, an acknowledgment of the conceptual priority of the concrete entities of classical mathematics, and a desire to improve our understanding of them, rather than a drive to encourage the study of mathematical entities defined by abstract axioms devoid of immediate, intuitive significance . . . All the concepts introduced in algebra derive their meaning, their justification and their properties from those of the systems of complex and real numbers, rather than the other way around ([4, p. 170]).

Here the axiomatic method is clearly separated from "abstract axioms." Corry claims that, for Hilbert, the axiomatic method had little to do with the abstract method and, what it had, it had only in a derivative manner. The key element, for Hilbert, is the desire to improve our understanding of concrete entities of classical mathematics, and the axiomatic method becomes a tool for that very purpose.

> For him [Hilbert], the real focus of interest lay in the interrelation among the various groups of axioms, rather than among the individual axioms across groups. For him, the groups corresponded to the isolable basis of our spatial intuition and the main task of his axiomatic approach was to show the way in which they logically interacted to create the body of geometric knowledge ([6, p. 5]).

Often, when a domain of inquiry is axiomatized, the goal is to lay bare the underlying logical structure of the system of definitions and proofs, because the field has become a complete conceptual mess. Clarity and order are brought by a clean and lean axiomatic development. The structure of logical dependence between (1) primitive notions and the defined ones are exhibited and (2) the axioms and the deduced propositions are displayed openly. Thus, a structure of justification between the notions as well as between the propositions and based solely on logical relationships becomes transparent. As such, the axioms can be presented as constituting the foundations of the given domain. Hilbert's work in the foundations of geometry belongs to that category.

Some philosophers and mathematicians attribute an additional epistemic function to axioms in this context: the axioms should provide basic, self-evident truths on which the whole given domain rests. Frege, for one, certainly believed that axioms had to have this epistemic feature.[7] This desideratum is clearly not an ingredient of the abstract method. It is hard to convince oneself that the axioms of group theory, for instance, are self-evident.

After Hilbert and using some of Hilbert's tools, logicians and mathematicians turned their attention to axioms themselves. This became known as postulational analysis and was energetically developed in the United States by E. H. Moore and his students.

> But in the case of E. H. Moore, his students at Chicago, and some other contemporary USA mathematicians, their study of the *Grundlagen* led to development of a point of view that diverged from Hilbert's in this significant yet subtle matter: they turned the analysis of systems of axioms into a field of intrinsic

> mathematical interest in which the requirements introduced by Hilbert oriented the research questions and afforded the main technical tools to deal with them ([6, p. 5]).

The function of the axiomatic method is now different. One wants to reduce the number of axioms, investigate their independence, the categoricity of the system, their (syntactic) coherence, and, later the semantic coherence, completeness, etc. Clearly, the abstract method has, in itself, nothing to do with this usage of axioms.

Of course, axioms and the axiomatic method did play a key role in the rise of the abstract method. The axioms capture the invariant features of the theories under investigation. After these have been identified, the axiomatic method allows for the systematic and rigorous development of the consequences of these features. One *could* use a different method of presentation of the invariant features. It depends on the linguistic means available. For instance, nowadays, it would be possible to use a graphical language to present a new theory by using what are called *sketches*. (See [1] for an introduction to the language of sketches as a method of presentation of theories.) Be that as it may, I think that the identification of the axiomatic method with the abstract method introduces some confusion. Indeed, some mathematicians started to attribute to the axiomatic method virtues that belong to the abstract method.[8] I will get back to this important point in the section on philosophical consequences.

Extracting a Criterion of Identity

I claim that the extraction of a criterion of identity has been the blind spot in the mathematicians' journey through the abstract method. Contemporary mathematicians nowadays suppose that the proper criterion of identity comes almost automatically with the right axiomatic theory. For this is how we are taught modern mathematics: one learns the axioms, some of their most important consequences, and, immediately after, the proper notion of homomorphism for the structure given and the resulting notion of isomorphism for that structure. We now take for granted the last notion and its importance. Historically, there is a time lag between the identification of a certain mathematical structure and its proper criterion of identity. This is not surprising as such. What is somewhat surprising is the length of the delay between the identification of the properties and the proper criterion of identity. Two historical cases can be mentioned: the case of homeomorphism for topological spaces and the case of equivalence for categories.

Topology slowly emerged as a field at the beginning of the twentieth century, and Hausdorff axiomatized the theory for the first time in 1914.[9] However, the definition of homeomorphism is nowhere to be found in the original edition of the book, although the concept is clearly there, probably under the influence of Fréchet's work, in which it is clearly identified. It is, however, identified as such by Hausdorff in his 1927 edition. Kuratowski wrote clearly and unambiguously in 1921 the clear conception of the notion of homeomorphism and its importance for topology, thus, almost ten years after Hausdorff's axiomatization. If one considers the roots of topology and looks at *analysis situs*, then one finds that it took far more than a decade to set the record straight; in fact, it took almost fifty years. According to G. H. Moore, "at different times and by different authors, at least four distinct concepts were identified in Euclidean spaces with those mappings under which topological properties were invariant" ([23, p. 333]). These four concepts are: the concept of injective continuous mapping, that of homeomorphism, the notion of deformation, and, finally, the concept of diffeomorphism. What will certainly surprise a contemporary mathematician is that "it took decades for mathematicians to learn to distinguish clearly between them" [23, p. 334]. These four different concepts were clearly seen as being different by the community at large by the 1930s.

Another significant example is provided by category theory. (See [16] and [18] for more on the history of category theory.) In Eilenberg and Mac Lane's original paper published in 1945 ([7]), one does find the notion of isomorphism of categories, and it was certainly assumed by Eilenberg and Mac Lane that the latter was the proper criterion of identity for categories. However, for reasons that would take us too far from our present concern, the notion of isomorphism of categories turned out to be inappropriate as a criterion of identity in the practice of category theory, in particular when one considers functor categories, for example, categories of sheaves in the context of algebraic geometry. I hasten to add that although Eilenberg and Mac Lane considered functor categories in their original article, these categories played a purely auxiliary role in the original paper. This fact changed completely when the theory came into the brains of Grothendieck and Kan in the mid-1950s. Indeed, functor categories moved to center stage from that moment on. Be that as it may, the proper criterion of identity for categories, namely, the notion of equivalence of categories, was introduced explicitly as such by Grothendieck in his famous *Tohoku* paper published in 1957 ([9]), twelve years after Eilenberg and MacLane's original paper.

Our main point is simple: it is not until the proper criterion of identity has been identified and applied systematically that the theory acquires an autonomy, both epistemological and ontological. Notice also that it is the presence of a new criterion of identity that allows us to say that we are indeed in the presence of a new type of abstraction, for as we have seen, the usage of the axiomatic method in itself does not entail the need for a new criterion of identity. In the case of a given classical domain with a given criterion of identity, the role of the axiomatic method is radically different than the one it plays in the context of the abstract method. The identification of the proper criterion of identity is of fundamental importance, since it allows us to sift the properties of the resulting theory from the properties of the previous theories. In other words, it captures the process of abstraction itself.

I claim that we now have all the pieces lying in front of us. I have presented the bare bones of the abstract method that characterized modern mathematics. We are now ready to see how the abstract method leads to levels of abstraction.

The Abstract Method and Levels of Abstraction

The goal in this section is to propose a more formal analysis of the notion of levels of abstraction in mathematics. I have to make two preliminary remarks. First, one could and perhaps should provide an analysis that would be completely formal, that is, one that would use logical systems and translations between logical theories to arrive at a precise, rigorous, and formal definition. I will leave this approach to another time. Second, I strongly believe that the following analysis illuminates certain aspects of the space of mathematical concepts, to use a metaphor, and even though it might turn out to be flawed, it is my conviction that it points in the correct direction. Third, I am *not* claiming that this analysis reflects in any way how these concepts were in fact discovered in the last century. I am here moving away from the historical record and considering various questions more from a logical point of view.

These remarks being made, let us start with a few paradigmatic cases of mathematical concepts that seem to be at different levels of abstraction.

I submit that the notion of topological space is more abstract than the concept of metric space.[10] Informally, the idea is that there is more significant variation among topological spaces than there is among metric spaces. This informal claim rests in part on a series of precise mathematical results that are well known.

First, every metric space give rises to a topological space in a canonical way. Second, every map between metric spaces, that is, every continuous map that does not increase any pairwise distance,[11] is trivially a continuous map between the associated topological spaces. In other words, there is a canonical functor from the category of metric spaces to the category of topological spaces. However, it is also a well-known result that not all topological spaces are metrizable. Only topological spaces satisfying certain properties are metrizable. Finally, two different metric spaces can give rise to the same topological space.

Are these facts enough to conclude that the concept of topological space is more abstract than the concept of metric space? Notice that both categories are "just as big," that is, it seems hopeless to exploit some cardinality condition in general. Being less abstract *cannot* merely be captured, for instance, by the fact that the extension of the concept D, for example, being a metric space, is strictly included in the concept C, for example, being a topological space. The structures involved, their properties, and how they are related have to be used to make sense of the notion of levels of abstraction. Notice also that it seems inappropriate to simply say that the notion of topological space is more general than the notion of metric space. It is indeed, but from an epistemological point of view, to focus on generality is to miss the point of the conceptual difference between the two notions.

Let us go up the stairs of abstraction. I claim that the notion of locale is more abstract than the notion of topological space. (See [13, 24] for more on the theory of locales.) The reasons are essentially the same as those of the foregoing example. To see this, let us briefly recall the definition of a locale and certain basic facts about them.

We first have to start with the algebraic notion of *frame*. A *frame* is a partially ordered set with finite meets and arbitrary joins such that meets distribute over joins. A morphism of frames is a map of sets that preserves finite meets and arbitrary joins. Notice that a frame is automatically a Heyting algebra, but that frame homomorphisms need not preserve the Heyting implication. In fact, frames are sometimes identified with complete Heyting algebras. We can therefore consider the category of frames and frame homomorphisms. The category of *locales* is the opposite of the category of frames, that is, it has the same objects but the morphisms go in the opposite direction.

To understand the link between locales and topological spaces, one has to consider the algebra of open sets of a topological space. It is easy to see that it is a frame. Thus, every topological space gives rise in a canonical way to a locale and every continuous map gives rise to a map of locales. However, the notion of locale, although closely related to the notion of topological space, yields a different mathematical theory of space. The crucial difference is that, in a very specific sense, the theory of locales is the theory of pointless topology. Whereas a topological space always has an underlying set of points, the basic notion in locale theory is the notion of open subspace, the notion of point being a special case of the latter. For instance, there are nonempty locales that are without points.[12] The key is of course the notion of point. In a category $\mathbb{C}$, a point of an object X of $\mathbb{C}$ is given by a morphism from the terminal object 1, assuming the latter exists, into X. In the category **Set** of sets and functions, there is a bijection between the elements of a set X and its points in this sense, for the terminal object of the category of sets is any singleton set. In the category **Top**, a point of a space X is a continuous morphism from the one-point space 1 into X. The locale corresponding to the one-point space 1 is of course the two-element poset, in fact, Boolean algebra **2** and, therefore, a point of a locale $\mathcal{O}(X)$ is a morphism $\mathbf{2} \rightarrow \mathcal{O}(X)$. Given that the category of locales is the opposite of the category of frames, points of locales correspond to frame homomorphisms $\mathcal{O}(X) \rightarrow \mathbf{2}$. As is well known, there are nonatomic Boolean algebras, thus pointless locales.

As in the previous case, there are different (nonsober) topological spaces, that is, nonhomeomorphic spaces that give rise to isomorphic locales.[13] Again, since there are pointless locales, the notion of locale is indeed also more general than the notion of topological space, and not all locales are spatial.[14] I claim that in this case also, it would be insufficient to merely focus on the difference in generality between the concepts. There is something else, something epistemologically more important at work.

Here is an example of two notions that I believe are at the same level of abstraction: the notion of group and the notion of abelian group. I submit that they are just as abstract and that the property that separates them does not introduce a difference in their level of abstraction.

It would be easy to multiply the examples and explore various cases. (I encourage the reader to do so!) But our goal here is more modest. We only presented these examples to convince the reader that it seems plausible to say that a given mathematical notion is more abstract than another, and I hope that these examples are plausible.

Let us sum up. We are now in the following position. First, there is circumstantial empirical evidence that mathematicians accept the idea that there are levels of abstraction in mathematics. Second, the abstract method, as it was developed in the twentieth century, clearly opens the door to a procedure that can yield different levels of abstraction, provided that the latter notion makes sense at all. Third, there are cases of mathematical notions about which it seems reasonable to say that they differ in their levels of abstraction. It is another matter to say exactly what differing in levels of abstraction might mean. Assuming that we have sufficient evidence to conclude that there is a need for some sort of philosophical clarification of the notion of levels of abstraction, I present the following proposal.

Let $\mathbb{T}_1$ and $\mathbb{T}_2$ be two mathematical theories such that there is a canonical way to transform every model of $\mathbb{T}_2$ into a model of $\mathbb{T}_1$. Given a model M of $\mathbb{T}_2$, we denote the canonical model of $\mathbb{T}_1$ it gives rise to by $\mathrm{F}(M)$. We say that $\mathbb{T}_1$ is *more abstract* than a theory $\mathbb{T}_2$ if

1. $(\forall M_1)(\forall M_2)(M_1 \cong_{\mathbb{T}_2} M_2 \Rightarrow F(M_1) \cong_{\mathbb{T}_1} F(M_2))$
2. $(\exists M_1)(\exists M_2)(F(M_2) \cong_{\mathbb{T}_1} F(M_1) \Lambda\, M \ncong_{\mathbb{T}_2} M_2)$..
3. The objects $F(M_2)$ and $F(M_1)$ are nontrivial.

The first clause simply says that the theories are connected in a systematic fashion, that is, the criteria of identity of $\mathbb{T}_2$ are preserved by the canonical transformation F. Notice that any functor between two categories of models automatically preserves isomorphisms. Hence, we have a specific framework in which this condition is easily satisfied.

More importantly, the functor F has to be *canonical*. This is obviously a tricky expression, and although it is used by mathematicians in numerous circumstances and occasions, it does not have a precise technical meaning. Nor will we try to give one to it.[15] The second condition captures the idea that we are dealing with an abstraction process. As we have indicated in the section entitled "Extracting a Criterion of Identity," criteria of identity allow one to determine the relevant properties of given objects. Thus, the condition stipulates that there is at least one property that allows one to distinguish between M_1 and M_2 but that when one moves to the canonical objects $F(M_1)$ and $F(M_2)$ arising from M_1 and M_2, respectively, it is no longer possible to distinguish them. In other words, the discriminating properties were abstracted from M_1 and M_2. The third condition is simply there to block trivial cases in which one would erroneously conclude that there is a new level of abstraction when, in fact, one has in some sense abstracted everything.

We can now assert that topology is more abstract than the theory of metric spaces and that the theory of locales is more abstract than topology. Our informal exposition earlier contains all the required ingredients to verify that the formal characterizations apply. I will leave to the reader the pleasure of exploring other cases and coming to its conclusions.

Some Philosophical Consequences

This section will be sketchy and brief. I will barely touch upon some of the consequences of the foregoing proposal here.

Epistemological Consequences of the Abstract Method

> The axiomatic method is thus first a method of economy of thought, in that it allows to condense many different reasonings in one. (...) But the axiomatic method is also a method of discovery and of progress.[16] (Weil, pp. 19–20 [32]) [my translation].

I believe this is another revealing case of a mathematician confusing the axiomatic method with the abstract method. I submit that it is really the abstract method that has the virtues expounded by Weil and that the axiomatic method is, in this particular context, a useful tool in the application and development of the abstract method.[17] Thus, one of the characteristics of the development of mathematics in the twentieth century is the systematic use of the abstract method and, as a consequence, the fact that mathematics has become more abstract in the foregoing precise sense. It is not the only aspect, but it is certainly one of the distinguishing features of modern mathematics. As Weil sees so clearly, the epistemological benefits of the abstract method are immense. First, many different results can now be proved in a uniform manner and from a common core. There is, to use contemporary language, a form of "compression" of the mathematical information. One and the same proof now applies to significantly different mathematical domains, and the proof exhibits the bare essentials involved in the result. This was also mentioned explicitly by Banach in the introduction of his book on functional analysis.

> The aim of the present work is to establish certain theorems valid in different functional domains, which I specify in what follows. Nevertheless, in order not to have to prove them for each particular domain, I have chosen to take a different route . . .; I consider sets of elements about which I postulate certain properties; I deduce from them certain theorems, and I then prove for each particular functional domain that the postulates adopted are true for it (Banach, quoted by Moore [22, p. 280]).

That is a crystal-clear exposition of the abstract method and its advantages.

Second, the abstract method is a method of discovery. Again, as such, the axiomatic method is not traditionally thought of in those terms.[18] Certainly, philosophers never thought of the axiomatic method as a method of discovery. In the context of the abstract method, however, it makes perfect sense. Indeed, given (at least) three domains where there is significant variation, if indeed one can discover common traits that can be used to prove important features of the situations, the method can be seen as a method of discovery and, as the quote by Weil shows, it has been considered as such.

Last, but not least, Weil associates the method with the idea of progress. I cannot venture into the conception of progress underlying Weil's claim, apart from the trivial claim that it is seen in a positive light. I will nonetheless speculate that progress in this particular context is linked to understanding. It can certainly be claimed that by identifying features that are abstracted from significantly different situations that nonetheless allow us to prove important facts of these situations, one is led to conclude that *essential* properties and relations have been brought to the fore and that these essential features, these new concepts, allow us to understand why these results hold. For now, we have the bare essentials involved in these significantly different cases. Notice that, as another benefit of the method, one obtains a form of unity of mathematics, a unity that is not so much the fact that all mathematical entities can be defined as sets, but rather as a working unity, as concepts that play important roles in different domains. Certain concepts arise in domains that are significantly different, sometimes in unexpected ways. It is of course their abstract character that makes this even possible.

Being Abstract

Traditionally, the property of being abstract is opposed to the property of being concrete. Thus, an entity is either abstract or concrete, period. As such, the distinction is presented as being an ontological distinction. The usual strategy is to provide a way to characterize what it is to be concrete, for example, being causally efficacious or having spatiotemporal coordinates, and then saying that something is abstract if it is not concrete. It is well known that these criteria all have shortcomings. No matter how you describe them, it is clear that mathematical entities are *not concrete* in the ontological sense. But I would rather oppose concrete to nonconcrete than to abstract. I imagine that it is possible for something to be not concrete without being abstract, that is, without being the result of a process of abstraction. Be that as it may, it should be clear by now that the problem I meant to explore is purely epistemological. It is rather a question of how to introduce differences *within* the realm of abstractions.

It is a consequence of my approach that the distinction between levels of abstraction is *relative to a given background*. I have not given an absolute characterization of levels of abstraction. In a sense, it is purely "local." For instance, the concept of topological space is more abstract than the concept of metric space, but it is less abstract than the concept of locale. Thus, although the distinction is not subjective, it is relative. Someone *can* say that something is abstract, whereas somebody else will say that the same piece of mathematics is not *that* abstract, and it won't constitute a contradiction. In these cases, one has to unearth the underlying context, the various domains of variation that are involved, the relationships between the concepts, if there is any, and the canonical connections between them, if there are any. It should also be obvious that there are concepts that are simply incomparable with respect to their level of abstraction. I believe this is as it should be.

Going Up the Ladder of Abstraction

I submit that mathematics in the twenty-first century *is*, as a matter of fact, becoming more abstract and it is an inevitable result of the application of the abstract method in various contexts.

Sometimes, mathematicians do not even see that they are introducing a domain of variation. This phenomenon happened at least twice in the twentieth century, and they turn out to be conceptually closely related: category theory and homotopy theory. I will ignore homotopy theory in this paper, although it certainly deserves a whole book. When they were first introduced by Eilenberg and Mac Lane, categories were not even considered as being genuine mathematical entities. They were merely conceived as a convenient tool. The language of category theory was thought to be merely useful in that it brought a certain order in fields that were at that time rather messy, for example, algebraic topology and the emerging homological algebra. But the introduction of a language introduces with it the possibility of expressing properties in a new way, expressing new properties, and detecting invariant patterns that were hidden before or simply nonexistent. If, moreover, that language has the resources to express in a new fashion criteria of identity, as is the case with category theory, then everything is in place for the process of abstraction to start and this is precisely what happened with category theory. First, categories include in an intrinsic fashion a criterion of identity for the objects of a given category, namely, the notion of isomorphism for that category. It naturally comes with the notion and the context. After you have identified a category, you have a notion of isomorphism, a criterion of identity, for its objects. However, as I have already mentioned, the proper criterion of identity for categories themselves is not the notion of isomorphism of categories. It is the notion of equivalence of categories.[19] It is easy to convince oneself that by adopting the latter criterion we are going up the ladder of abstraction. And it does not stop there. In fact, there is a whole hierarchy of levels, called higher-dimensional categories, with each new level being more abstract than the preceding. Presenting the theory of weak n-categories would require too much space and would lead us into unnecessary technical complexities for our purposes. Suffice it to say that our framework is coherent with the claim that contemporary mathematics is becoming more abstract. The foregoing framework allows us to say what it could mean.

Needless to say, what goes up must go down, and this is just as important in mathematics. Abstracting is often accompanied with representations. In fact, one can analyze representation theory for groups and other concepts and a systematic interplay between the abstract and the concrete and how to use one level of knowledge to illuminate another level. Again, a more comprehensive analysis would be required and will be left for another paper.

Abstracting as a Way of Simplifying

The abstract method constituted a clear cognitive gain. As I have indicated, going in the direction of abstraction is a way of unifying different mathematical domains. The introduction of a level of abstraction is seen as a way of clarifying and distilling what, in some cases, has become a complex domain or, in other cases, exhibits similarities, parallels indicating the possibility of an underlying common framework. The previous disjunction is clearly not exclusive. The new abstract level not only simplifies the situation but it also yields a better control and understanding of the concepts involved. The axiomatic method is an essential part of that process. Axiomatization should be seen, in this light, as a form of design. For instance, in the case of analogies or similarities, axioms capture either a common structure or common properties, again leading to a better control and understanding of the features at work. The abstract method is thus used as a seam, a filter in these processes. It brings to the fore the Archimedean points on which solutions to given problems work. What was previously immersed in a mountain of irrelevant details is unearthed and shown to constitute the mechanisms making concepts work together. This is precisely why we feel justified in speaking of abstraction. As I have said, the process leads to new mathematics, conceptually systematic and organized according to clear principles.

In recent times, the abstract method has even been presented as inevitable. Here is a wonderful quote illustrating this claim:

> The first part of the paper, on which everything else depends, may perhaps look a little frightening because of the abstract language that it uses throughout. This is unfortunate, but there is no way out. It is not the purpose of the abstract language to strive for great generality. The purpose is rather to simplify proofs, and indeed to make some proofs understandable at all. The reader is invited to run the following test: take theorem 2.2.1 (this is about the worst case), translate the complete proof not using the abstract language, and then try to communicate it to somebody else [29, p. 318].

It is not possible to go into Waldhausen's remarkable and fundamental paper here, which provides an abstract and flexible setting for algebraic K-theory. We have to take his claims as a reliable testimony in the present context. But this brief paragraph contains important and bold claims that would deserve to be unpacked properly. Waldhausen insists that his approach is *not* motivated by the search for the most general solution. He was aware that his framework was more general than Quillen's framework for algebraic K-theory introduced in the previous decade. Generality is merely a by-product of the search for the right context. The fundamental purpose is "to make some proofs understandable at all." The latter is achieved by a process of simplification, which, in turn, is itself a by-product of the abstract method.

Understanding how there can be different levels of abstraction also helps us to see how there can be different mathematical styles. For instance, using categories right from the start and trying to solve given mathematical problems by using categorical concepts and methods and finding a solution up to categorical equivalence is a more abstract way of doing mathematics than by using sets with structures and the resulting notion of isomorphism involved. The reader who knows even just a little bit about the history of algebraic geometry in the second half of the twentieth century can recognize Grothendieck's style in the former and why it was seen as being so outlandish by many. Grothendieck thought that by going up the ladder, one would see better the overall organization of the field, would understand better the general topography of the landscape, and that mathematicians would therefore find their way more easily. In his mind, the path to the proof would be trivial and obvious. Abstraction is seen as a radical way to simplify mathematics. Of course, this is possible only after one has understood how to go up the ladder and that it is a ladder. Others find these heights dizzying or consider that going that far up is not worth the effort.

Notes

1. I urge readers to consult the paper by Hourya Benis Sinaceur for a slightly different point of view on the matter. See [2].

2. The temporal order of these moments can vary. Indeed, historically one finds various combinations, although most cases seem to follow the order of our presentation, for reasons that are not hard to understand. It is, however, interesting to note that there are cases in which the development follows a different path.

3. It should be obvious to the reader that the former conceptual analysis can be developed and used in different historical contexts. My goal in this paper is to focus on one particular episode, namely, the rise of what is called modern mathematics. For the latter particular purpose, I feel justified in using a more formal language than might be adequate for other periods where abstraction is at work.

4. It is known that Fréchet knew about the case of groups and that it provided at least guidelines and a model of what could be achieved by moving up the ladder of abstraction.

5. I have to come back to my earlier remark concerning the temporal order involved here. It is tempting to read the diagram from left to right and to think that there is a chronological order in that reading. Although it is in most cases correct, there are counterexamples to that reading, e.g., Boolean algebras or lattices. What seems to be correct is the step from the presentation of the theory to the extraction of the criterion of identity. As far as I know, that step is always taken in that order, although after the method has been well understood, the delay between the presentation of the theory and the extraction of the criterion of identity can be very short.

6. A referee pointed out to me that it is certainly historically abusive to talk about theories when referring to the domains that led to the birth of abstract group theory. Clearly, algebraic equations, number theory, and geometry were not formal axiomatized theories in the eighteenth and nineteenth centuries. The point is well taken. The reader might want to think about certain languages and a set of problems about disparate kinds of mathematical things. I believe that my conceptual analysis would still work.

7. Needless to say, this epistemic standpoint was not and is certainly not adopted by all mathematicians and logicians using the axiomatic method this way. In fact, the whole spectrum of positions is likely to be found within the community of mathematicians and logicians. Contrast, for instance, the following claim made by Hölder:

> [the axioms of arithmetic] which are based on the fact that we take it as evident that there are certain processes, which, as we say, proceed according to determinate rules that are always performable in a definite way and in certain cases can be carried out repeatedly without end (Hölder, quoted by Scanlan [25, p. 988]).

where Hölder mentions that we take certain things or processes as being evident, with the following claim made by his contemporary Peano

> Logical questions thus become completely independent of *empirical or psychological* questions (and, in particular, the *problem of knowledge*) and every question concerning the *simplicity of ideas* and the *obviousness of facts* disappears (Peano, quoted by Scanlan [25, p. 988]).

The distance between these two positions could hardly be greater.

8. A referee rightly pointed out that one could compare and contrast the abstract method with the method of analysis. The latter has a long and convoluted history, but certainly occupies a central role in the development of mathematics. It could be argued that the abstract method that I describe here is but one special instance of the method of analysis. This point would, however, require a whole paper to be explained and argued properly.

9. The axiomatization of topology went through various phases from 1914 until the late 1950s. This is an interesting case in which the different axiomatic frameworks are proposed mainly for practical reasons, that is, the modifications suited the specific needs of various mathematicians at different times.

10. I guess one could conduct empirical research among professional mathematicians and ask them whether they would agree or not with claims of the sort: "the concept of topological space is more abstract than the concept of metric space." I have not conducted such research. I have used claims found in the literature that illustrate the fact that mathematicians talk that way.

11. The definition of maps between metric spaces is always a delicate matter. We have chosen a simple class of morphisms. The main point remains, no matter how we choose them.

12. Thus, it seems rather obvious that the notion of locale could be useful for mereology. As far as I know, no one has looked at the links between traditional mereological notions and the theory of locales.

13. This is not an entirely trivial result. It requires some fiddling around. One way to build the appropriate spaces can be found in [3].

14. Informally, a locale is spatial if it arises from the lattice of open sets of a topological space. It is of course possible to give an intrinsic characterization.

15. In practice, mathematicians know what it means to say that something is given canonically or that there is a canonical construction or presentation of something. Very roughly, it means that the construction does not depend on a choice from a mathematician in the process. Whether this can be made more precise is an open question. See, however, [19] on the subject

16. "La méthode axiomatique est donc d'abord méthode d'économie de pensée, en ceci qu'elle permet de condenser en un seul plusieurs raisonnements différents. (...) Mais la méthode axiomatique est aussi méthode de découverte et de progrès."

17. As should be clear by now, the main point here is that the abstract method and the axiomatic method are distinct. They can interact in various ways, and they have. I want to thank an anonymous referee who forced me to clarify these points.

18. See, however, the paper [26].

19. For a nice discussion surrounding the notion of equality in modern mathematics, see [21].

References

[1] Michael Barr and Charles Wells. "Category theory for computing science." *Repr. Theory Appl. Categ.*, (22):xviii+538, 2012.

[2] Hourya Benis Sinaceur. "Facts and levels of mathematical abstraction." *Philosophia Scientiae*, 18(1):1–32, 2014.

[3] Olivia Caramello. "A topos-theoretic approach to Stone-type dualities." 03 2011. Available on the math arxiv: arXiv:math.CT/1006.3930.

[4] Leo Corry. *Modern algebra and the rise of mathematical structures*, volume 17 of *Science Networks. Historical Studies*. Birkhäuser Verlag, Basel, Switzerland, 1996.

[5] Leo Corry. "The origins of the definition of abstract rings." *Gaz. Math.*, (83):29–47, 2000.

[6] Leo Corry. "Axiomatics between Hilbert and the new math: Diverging views on mathematical research and their consequences on education." *International Journal for the History of Mathematics Education*, 2(2):21–37, 2007.

[7] Samuel Eilenberg and Saunders Mac Lane. "A general theory of natural equivalences." *Trans. Amer. Math. Soc.*, 58:231–294, 1945.

[8] Maurice Fréchet. "Sur quelques points du calcul fonctionnel." *Rendiconti Circ. Mat. Palermo*, 22:1–74, 1906.

[9] Alexander Grothendieck. "Sur quelques points d'algèbre homologique." *Tôhoku Math. J. (2)*, 9:119–221, 1957.

[10] Helmut Hasse. "The modern algebraic method." *Math. Intelligencer*, 8(2):18–25, 1986. Reprinting of *Jahresber. Deutsch. Math.-Verein.* **30** (1930), 22–34, Translated from the German by Abe Shenitzer, with comments by Bruce Chandler and Wilhelm Magnus.

[11] I. N. Herstein. *Topics in algebra,* 2nd ed., Xerox College Publishing, Lexington, MA, 1975.

[12] Nathan Jacobson. *Basic algebra. I.* W. H. Freeman and Co., San Francisco, 1974.

[13] Peter T. Johnstone. *Stone spaces*, volume 3 of *Cambridge Studies in Advanced Mathematics*. Cambridge University Press, Cambridge, U.K., 1986. Reprint of the 1982 edition.

[14] Israel Kleiner. "The genesis of the abstract ring concept." *Amer. Math. Monthly*, 103(5):417–424, 1996.

[15] Israel Kleiner. *A history of abstract algebra*. Birkhäuser Boston Inc., Boston, 2007.

[16] Ralf Krömer. *Tool and object*, volume 32 of *Science Networks. Historical Studies*. Birkhäuser Verlag, Basel, Switzerland, 2007. A history and philosophy of category theory.

[17] Saunders MacLane. "Some recent advances in algebra." *Amer. Math. Monthly*, 46(1):3–19, 1939.

[18] Jean-Pierre Marquis. *From a geometric point of view: A study in the history and philosophy of category theory*, volume 14 of *Logic, Epistemology, and the Unity of Science*. Springer, New York, 2009.

[19] Jean-Pierre Marquis. "Canonical maps" to appear in the book *Categories for the working philosopher*, Elaine Landry, ed., Oxford University Press. Forthcoming.

[20] Jean-Pierre Marquis. "Mathematical abstraction, conceptual variation and identity." In P.-E. Bour; G. Heinzmann; W. Hodges; P. Schroeder-Heister, editors, *Logic, Methodology and Philosophy of Science, proceedings of the fourteenth international congress*, College Publications, London, 2015, pp. 299–322.

[21] Barry Mazur. "When is one thing equal to some other thing?" In *Proof and other dilemmas*, MAA Spectrum, Math. Assoc. America, Washington, DC, 2008, pp. 221–241.

[22] Gregory H. Moore. "The axiomatization of linear algebra: 1875–1940." *Historia Math.*, 22(3):262–303, 1995.

[23] Gregory H. Moore. "The evolution of the concept of homeomorphism." *Historia Math.*, 34(3):333–343, 2007.

[24] Jorge Picado and Aleš Pultr. *Frames and locales*. Frontiers in Mathematics. Birkhäuser/Springer Basel AG, Basel, Switzerland, 2012. Topology without points.

[25] Michael Scanlan. "Who were the American postulate theorists?" *J. Symbolic Logic*, 56(3):981–1002, 1991.

[26] Dirk Schlimm. "Axioms in mathematical practice." *Philos. Math. (3)*, 21(1):37–92, 2013.
[27] R. Solomon. *Abstract Algebra*. Pure and applied undergraduate texts. American Mathematical Society, Providence, RI, 2003.
[28] Angus E. Taylor. "A study of Maurice Fréchet. I. His early work on point set theory and the theory of functionals." *Arch. Hist. Exact Sci.*, 27(3):233–295, 1982.
[29] Friedhelm Waldhausen. "Algebraic *K*-theory of spaces." In *Algebraic and geometric topology (New Brunswick, N.J., 1983)*, volume 1126 of *Lecture Notes in Math.*, pp. 318–419. Springer, Berlin, 1985.
[30] John J. Watkins. *Topics in commutative ring theory*. Princeton University Press, Princeton, NJ, 2007.
[31] Hans Wussing. *The genesis of the abstract group concept*. MIT Press, Cambridge, MA, 1984. A contribution to the history of the origin of abstract group theory, Translated from the German by Abe Shenitzer and Hardy Grant.
[32] André Weil. "Livre 1. Theorie des ensembles. Introduction." http://sites.mathdoc.fr/archives-bourbaki/PDF/066_awr_002.pdf.

Are Our Brains Bayesian?

Robert Bain

The human brain is made up of 90 billion neurons connected by more than 100 trillion synapses. It has been described as the most complicated thing in the world, but brain scientists say that is wrong: they think it is the most complicated thing in the known universe. Little wonder, then, that scientists have such trouble working out how our brain actually works. Not in a mechanical sense: we know, roughly speaking, how different areas of the brain control different aspects of our bodies and our emotions, and how these distinct regions interact. The questions that are more difficult to answer relate to the complex decision-making processes each of us experiences: how do we form beliefs, assess evidence, make judgments, and decide on a course of action?

Figuring that out would be a great achievement, in and of itself. But this has practical applications, too, not least for those artificial intelligence (AI) researchers who are looking to transpose the subtlety and adaptability of human thought from biological "wetware" to computing hardware.

In looking to replicate aspects of human cognition, AI researchers have made use of algorithms that learn from data through a process known as Bayesian inference. Based on Bayes' famous theorem (see box called Bayes' Theorem), Bayesian inference is a method of updating beliefs in the light of new evidence, with the strength of those beliefs captured using probabilities. As such, it differs from frequentist inference, which focuses on how frequently we might expect to observe a given set of events under specific conditions.

Bayesian inference is seen as an optimal way of assessing evidence and judging probability in real-world situations—"optimal" in the sense that it is ideally rational in the way it integrates different sources of information to arrive at an output. In the field of AI, Bayesian inference

Bayes' Theorem

Bayes' theorem provides us with the means to calculate the probability of certain events occurring conditional on other events that may occur—expressed as a probability between 0 and 1. Consider the following formula:

$$P(B|E) = \frac{P(E|B)\,P(B)}{P(E)}$$

This tells us the probability of event B occurring given that event E has happened. This is known as a conditional probability, and it is derived by multiplying the conditional probability of E given B, by the probability of event B, divided by the probability of event E.

The same basic idea can be applied to beliefs. In this context, $P(B|E)$ is interpreted as the strength of belief B given evidence E, and $P(B)$ is our prior level of belief before we came across evidence E.

Using Bayes' theorem, we can turn our "prior belief" into a "posterior belief." When new evidence arises, we can repeat the calculation, with our last posterior belief becoming our next prior. The more evidence we assess, the sharper our judgments should become.

has been found to be effective at helping machines approximate some human abilities, such as image recognition. But are there grounds for believing that this is how human thought processes work more generally? Do our beliefs, judgments, and decisions follow the rules of Bayesian inference?

Beneath the Surface

For the clearest evidence of Bayesian reasoning in the brain, we must look past the high-level cognitive processes that govern how we think and assess evidence, and consider the unconscious processes that control perception and movement.

Professor Daniel Wolpert of the University of Cambridge's neuroscience research center believes that we have our Bayesian brains to thank for allowing us to move our bodies gracefully and efficiently—by making reliable, quick-fire predictions about the result of every movement we make.

Imagine taking a shot at a basketball hoop. In the context of a Bayesian formula, the "belief" would be what our brains already know about the nature of the world (how gravity works, how balls behave, every shot we have ever taken), while the "evidence" is our sensory input about what is going on right now (whether there is any breeze, the distance to the hoop, how tired we are, the defensive abilities of the opposing players).

Wolpert, who has conducted a number of studies on how people control their movements, believes that as we go through life, our brains gather statistics for different movement tasks and combine these in a Bayesian fashion with sensory data, together with estimates of the reliability of those data. "We really are Bayesian inference machines," he says.

To demonstrate this, Wolpert and his team invited research subjects to their lab to undergo tests based on a cutting-edge neuroscientific technique: tickling.

As every five-year-old knows, other people can easily have you in fits of laughter by tickling you, but you cannot tickle yourself. Until now. Wolpert's team used a robot arm to mediate and delay people's self-tickling movements by fractions of a second—which turned out to be enough to bring the tickly feeling back.

The problem with trying to tickle yourself is that your body is so good at predicting the results of your movement and reacting that it cancels out the effect. But by delaying people's movements, Wolpert was able to mess with the brain's predictions just enough to bring back the element of surprise. This revealed the brain's highly attuned estimates of what will happen when you move your finger in a certain way, which are similar to what a Bayesian calculation would produce from the same data.[1]

A study by Erno Téglás et al., published in *Science*, looked at one-year-old babies and how they learn to predict the movement of objects—measured by how long they looked at the objects as they moved them around in different patterns. The study found that the babies developed "expectations consistent with a Bayesian observer model," mirroring "rational probabilistic expectations."[2] In other words, the babies learned to expect certain patterns of movement, and their expectations were consistent with a Bayesian analysis of the probability of different patterns occurring.

Other researchers have found indications of Bayesianism in higher level cognition. A 2006 study by Tom Griffiths of the University of California, Berkeley, and Josh Tenenbaum of MIT asked people to make predictions of how long people would live, how much money films would make, and how long politicians would last in office. The only data they were given to work with was the running total so far: current age, money made so far, and years served in office to date. People's predictions, the researchers found, were very close to those derived from Bayesian calculations.[3]

This suggests that the brain not only has mastered Bayes' theorem, but also has finely tuned prior beliefs about these real-life phenomena, based on an understanding of the different distribution patterns of human life spans, box office takings, and political tenure. This is one of a number of studies that provide evidence of probabilistic models underlying the way we learn and think.

However, the route the brain takes to reach such apparently Bayesian conclusions is not always obvious. The authors of the study involving the babies said it was "unclear how exactly the workings of our model correspond to the mechanisms of infant cognition," but that the close fit between the model and the observed data about the babies' predictions suggests "at least a qualitative similarity between the two."[2]

To achieve such Bayesian reasoning, the brain would have to develop some kind of algorithm, manifested in patterns of neural connections, to represent or approximate Bayes' theorem. Eric-Jan Wagenmakers, an experimental psychologist at the University of Amsterdam, says the nature of the brain makes it difficult to imagine something as neat and elegant as Bayes' theorem being reflected in a form we would recognize. We are dealing, he says, with "an unsupervised network of very stupid individual units that is forced by feedback from its environment to mimic a system that is Bayesian."

Probability Puzzles

Before we accept the Bayesian brain hypothesis wholeheartedly, there are a number of strong counterarguments that must be considered. For starters, it is fairly easy to come up with probability puzzles that should yield to Bayesian methods, but that regularly leave many people flummoxed. For instance, many people will tell you that if you toss a series of coins, getting all heads or all tails is less likely than getting, for instance, tails–tails–heads–tails–heads. It is not, and Bayes' theorem shows why: as the coin tosses are independent, there is no reason to expect that one sequence is more likely than another.

There is also the well-known Monty Hall problem, which can fool even trained minds. Participants are asked to pick one of three doors—A, B, or C—behind one of which lies a prize. The host of the game then opens one of the non-winning doors (say, C), after which the contestant is given the choice of whether to stick with their original door (say, A) or switch to the other unopened door (say, B). Long story short: most people think switching will make no difference, when in fact it improves your chances of winning. Mathematician Keith Devlin has shown how you can use Bayes' formula to work this out, so that the prior probability that the prize is behind door B, 1 in 3, becomes 2 in 3 when door C is opened to reveal no prize. The details are online (bit.ly/montyhall1205) and they are fairly straightforward. Surely a Bayesian brain would be perfectly placed to cope with calculations such as these?

Even Sir David Spiegelhalter, professor of the public understanding of risk at the University of Cambridge, admits to mistrusting his intuition when it comes to probability. "The only gut feeling I have about probability is not to trust my gut feelings, and so whenever someone throws a problem at me I need to excuse myself, sit quietly muttering to myself for a while, and finally return with what may or may not be the right answer," Spiegelhalter writes in guidance for maths students (nrich.maths.org/7326).

Psychologists have uncovered plenty of examples where our brains fail to weigh up probabilities correctly. The work of Nobel Prize winner and bestselling author Daniel Kahneman, among others, has thrown light on the strange quirks of how we think and act—yielding countless examples of biases and mental shortcuts that produce questionable decisions. For instance, we are more likely to notice and believe information if it confirms our existing beliefs. We consistently assign too much weight to the first piece of evidence we encounter on a subject. We overestimate the likelihood of events that are easy to remember—which means that the more unusual something is, the more likely our brains think it is to happen.

These weird mental habits are a world away from what you would expect of a Bayesian brain. At the very least, they suggest that our "prior beliefs" are hopelessly skewed.

"There's considerable evidence that most people are dismally non-Bayesian when performing reasoning," says Robert Matthews of Aston University, Birmingham, and author of *Chancing It*, about the challenges of probabilistic reasoning. "For example, people typically ignore base-rate effects and overlook the need to know both false positive and false negative rates when assessing predictive or diagnostic tests."

In the case of testing for diseases, Bayes' theorem reveals how even a test that is said to be 99% accurate might be wrong half of the time when telling people they have a rare condition—because its prevalence is so low that even this supposedly high accuracy figure still leads to just as many false positives as true positives. It is a counterintuitive result that would elude most people not equipped with a calculator and a working knowledge of Bayes' theorem (see box called Diagnostic Test Accuracy Explained).

As for base rates, Bayes' theorem tells us to take these into account as "prior beliefs"—a crucial step that our mortal brains routinely overlook when judging probabilities on the fly.

All in all, that is quite a bit of evidence in favor of the argument that our brains are non-Bayesian. But do not forget that we are dealing with the most complicated thing in the known universe, and these fascinating quirks and imperfections do not give a complete picture of how we think.

Eric Mandelbaum, a philosopher and cognitive scientist at the City University of New York's Baruch College, says this kind of irrationality "is most striking because it arises against a backdrop of our extreme competence. For every heuristics-and-biases study that shows that we, for instance, cannot update base rates correctly, one can find instances where people do update correctly."

Tom Griffiths recently co-wrote the book *Algorithms to Live By* with Brian Christian. The authors agree that humans need to give themselves

Diagnostic Test Accuracy Explained

How is it that a diagnostic test that claims to be 99% accurate can still give a wrong diagnosis 50% of the time? In testing for a rare condition, we scan 10,000 people. Only 1% (100 people) have the condition; 9,900 do not. Of the 100 people who do have the disease, a 99% accurate test will detect 99 of the true cases, leaving one false negative. But a 99% accurate test will also produce false positives at the rate of 1%. So, of the 9,900 people who do not have the condition, 1% (99 people) will be told erroneously that they do have it. The total number of positive tests is therefore 198, of which only half are genuine. Thus the probability that a positive test result from this "99% accurate" test is a true positive is only 50%.

more credit for how well they make decisions. They write, "Over the past decade or two, behavioural economics has told a very particular story about human beings: that we are irrational and error-prone, owing in large part to the buggy, idiosyncratic hardware of the brain. This self-deprecating story has become increasingly familiar, but certain questions remain vexing. Why are four-year-olds, for instance, still better than million-dollar supercomputers at a host of cognitive tasks, including vision, language, and causal reasoning?"[4]

So while our well-documented flaws may shed light on the limits of our capacity for probabilistic analysis, we should not write off the brain's statistical abilities just yet. Perhaps what our failings really reveal, Griffiths and Christian suggest, is that life is full of really hard problems, which our brains must try to solve in a state of uncertainty and constant change, with scant information and no time.

Critical Thinking

Bayesian models of cognition remain a hotly debated area. Critics complain of too much *post hoc* rationalization, with researchers tweaking their models, priors, and assumptions to make almost any results fit a probabilistic interpretation. They warn of the Bayesian brain theory becoming a one-size-fits-all explanation for human cognition.

In a review of past studies on Bayesian cognition, Gary Marcus and Ernest Davis of New York University say, "If the probabilistic approach is to make a lasting contribution to researchers' understanding of the mind, beyond merely flagging the obvious facts that people are sensitive to probabilities and adjust their beliefs (sometimes) in light of evidence, its practitioners must face apparently conflicting data with considerably more rigour. They must also reach a consensus on how models will be chosen, and stick to that consensus consistently."[5]

Mandelbaum also has doubts about how far the Bayesian brain theory can go when it comes to higher level cognition. For a theory that purports to deal with updating beliefs, it struggles to explain the vagaries of belief acquisition, for instance. "Good objections to our views leave us unmoved," says Mandelbaum. "Excellent objections cause us to be even more convinced of our views. It is the rare egoless soul who finds arguments persuasive for domains they care deeply about. It is somewhat ironic that the area for which Bayesianism seems most well suited—belief updating—is the area where Bayesianism has the most problems."

Staying Alive

Stepping back from studies of the brain and looking at the bigger picture, some would take the fact that humanity is still here as evidence that something Bayesian is going on in our heads. After all, Bayesian decision-making is essentially about combining existing knowledge and new evidence in an optimal fashion. Or, as the renowned French mathematician Pierre-Simon Laplace put it, it is "common sense expressed in numbers." Wagenmakers says, "It's difficult to believe that any organism isn't at least approximately Bayesian in its environment, given its limits. Even if people use a heuristic that's not Bayesian, you could argue it's Bayesian because of the utility argument: the effort you have to put in is easy. Many decisions need to be fast decisions, so you have to make shortcuts sometimes. The shortcut itself is sub-optimal and non-Bayesian—but the fact you're taking the shortcut every time means huge savings in effort and energy. This may boil down to a more successful organism that's more Bayesian than its competitor."

On the other hand, having a Bayesian brain can still lead us into trouble. In a 2005 article in *Significance*, Robert Matthews explained how Bayesian inference is powerless in the face of strongly held irrational beliefs such as conspiracy theories and psychiatric delusions—because people's trust in new evidence is so low.[6] In this way, the mechanism by which a Bayesian brain updates beliefs can become corrupted, leading to false beliefs becoming reinforced by fresh evidence, rather than corrected.

To be fair, not even the most fervent supporter of the Bayesian brain hypothesis would claim that the brain is a perfect Bayesian inference engine. But we could reasonably describe a brain as "Bayesian" if it were "approximately optimal given its fundamental limitations," says Wagenmakers. "For example, people do not have infinite memory or infinite attentional capacity."

Another crucial caveat, Wagenmakers says, is that the Bayesian aspects of our brains are rooted in the environment in which we operate. "It's approximately Bayesian given the hardware limitations and given that we operate in this particular world. In a different world we might fail completely. We would not be able to adjust."

In their book, Christian and Griffiths argue that much of what the brain does is about making trade-offs in difficult situations—rather than aiming for perfection, the challenge becomes "how to manage finite space, finite time, limited attention, unknown unknowns, incomplete information, and an unforeseeable future; how to do so with grace and confidence; and how to do so in a community with others who are all simultaneously trying to do the same."[4]

Plenty to Learn

Clearly the Bayesian brain debate is far from settled. Some things our brains do seem to be Bayesian, others are miles off the mark. And there is much more to learn about what is really happening behind the scenes.

In any case, if our Bayesian powers are limited by the size of our brains, the time available, and all the other constraints of the environment in which we live (which differ from person to person and from moment to moment), then judging whether what is going on is Bayesian quickly becomes impossible, says Wagenmakers. "At some point it becomes a philosophical discussion whether you would call that Bayesian or not."

Even for those who are not convinced by the Bayesian brain hypothesis, the concept is proving a useful starting point for research.

Mandelbaum says: "At this point, it's more an idea guiding a research programme than a specific hypothesis about how cognition works."

Bayesian brain theories are used as part of rational analysis, which involves developing models of cognition based on a starting assumption of rationality, seeing whether they work, then reviewing them. Tom Griffiths says, "It turns out using this approach for making models of cognition works quite well. Even though there are ways that we deviate from Bayesian inference, the basic ideas behind it are right."

So the idea of the Bayesian brain is a gift to researchers because it "gives us a tool for understanding what the brain *should* be doing," says Griffiths. And for a lot of what the brain does, Bayesian inference remains "the best game in town in terms of a characterisation of what that ideal solution looks like."

References

Unreferenced quotations from Tom Griffiths, Eric Mandelbaum, Robert Matthews and Eric-Jan Wagenmakers are from interviews conducted by the author.

1. Blakemore, S. J., Wolpert, D., and Frith, C. (2000) "Why can't you tickle yourself?" *NeuroReport*, **11**(11), R11–16.
2. Téglás, E., Vul, E., Girotto, V., Gonzalez, M., Tenenbaum, J. B., and Bonatti, L. L. (2001) "Pure reasoning in 12-month-old infants as probabilistic inference." *Science*, **332**(6033), 1054–1059.
3. Griffiths, T. L., and Tenenbaum, J. B. (2006) "Optimal predictions in everyday cognition." *Psychological Science*, **17**(9), 767–773.
4. Christian, B., and Griffiths, T. (2016) *Algorithms to Live By*. London: William Collins.
5. Marcus, G. F., and Davis, E. (2013) "How robust are probabilistic models of higher-level cognition?" *Psychological Science*, **24**(12), 2351–2360.
6. Matthews, R. (2005) "Why do people believe weird things?" *Significance*, **2**(4), 182–185.

Great Expectations: The Past, Present, and Future of Prediction

Graham Southorn

We are all fascinated by the future. Whether it is the rise and fall in interest rates, the outcome of elections, or winners at the Oscars, there is sure to be something you want to know ahead of time. There is certainly no shortage of pundits with ready opinions about what the future might hold—but their predictions might not be entirely reliable. A 20-year study, published in 2006, showed that the average expert did little better than guessing on many of the political and economic questions asked of them.[1]

But expert predictions are only part of the forecasting story (see "Prediction versus Forecasting"). A raft of methods—from mathematical models to betting markets—are promising new ways of seeing into the future. And it is not only academics and professionals who can do it—online services allow anyone to have a go.

Forecasting could probably stake a claim to being one of the world's oldest professions. Beginning in the eighth century BC, a priestess known as the Pythia would answer questions about the future at the Temple of Apollo on Greece's Mount Parnassus.[2] It is said that she, the Oracle of Delphi, dispensed her wisdom in a trance—caused, some believe, by the hallucinogenic gases that would seep up through natural vents in the rock.

By the second century BC, the ancient Greeks had moved on to more sophisticated methods of prediction, such as the Antikythera mechanism, whose intricate bronze gears seemed capable of predicting a host of astronomical events such as eclipses.

Over time, our astronomical predictions became more refined, and in 1687 Isaac Newton published his laws of motion and gravitation.

We Need to Talk about Nate

One name crops up alongside virtually every mention of forecasting. Nate Silver is the high-profile American statistician best known for developing a mathematical model that correctly called the results in 49 of 50 states in the 2008 presidential election. The clever part was the way it incorporated polling data, explains Professor Rob Hyndman. "At least for the last two presidential elections, Nate Silver developed some very good Bayesian techniques for combining all of the prediction polls to get good forecasts of what would happen on election day."

Bayesian techniques are rooted in Bayes' theorem, which provides a means of updating levels of belief in the light of new evidence. Thus, as new polls are taken, Bayesian methods allow their findings to be combined with current polling evidence to produce an updated level of belief about, say, the various outcomes of an election.

But Silver's foresight may not be as startling as it first appears. In his book *Superforecasting*, Philip Tetlock points out that a "no change" forecast, in which the political party that won a state in the previous election merely holds onto it, would have correctly predicted 48 out of 50 results.

The same idea in meteorology is known as persistence forecasting. That is, the weather in future is forecast to be the same as it is now. The technique is only useful for very short-range forecasts or slowly evolving weather patterns.

Finally, it is worth noting that a forecast is only as good as the data it is based on. In the 2015 U.K. general election, Silver forecast that the Conservative Party would win more seats than Labour—but not as many as the Conservatives actually gained: the model's prediction interval fell short of the final tally of 331 seats. In this case, the opinion polls used by Silver, and many others, failed to accurately reflect voting intentions, which had a knock-on effect, or cumulative effect, on prediction accuracy.

Prediction versus Forecasting

The terms "prediction" and "forecasting" are often used interchangeably—as is the case in this article. But as far as anyone has managed to pin down a definition, one school of thought holds that forecasting is about the future—tomorrow's temperature, for example. Prediction, in contrast, involves finding out about the unobserved present. If you want to determine how much your house will sell for, you could make a prediction based on the prices of houses in your neighborhood.

In a blog post on this issue (bit.ly/21vlCGI), Galit Shmueli, Distinguished Professor of Business Analytics at Taiwan's National Tsing Hua University, wrote, "The term 'forecasting' is used when it is a time series and we are predicting the series into the future. Hence 'business forecasts' and 'weather forecasts.' In contrast, 'prediction' is the act of predicting in a cross-sectional setting, where the data are a snapshot in time (say, a one-time sample from a customer database). Here you use information on a sample of records to predict the value of other records (which can be a value that will be observed in the future)."

Newton's friend Edmond Halley predicted in 1705 the return of the comet that now bears his name. But forecasters also started to concern themselves with more mundane, earthly matters. By the nineteenth century, the new technology of long-distance telegrams meant that, for the first time, data from a network of weather stations could be transmitted in advance of changing conditions. This development not only spurred developments in meteorology. People also began to believe that similarly scientific measurements might be useful in other areas, such as business.

Forerunners

Early economic forecasters, such as Roger Babson, have been profiled by Walter A. Friedman in his book *Fortune Tellers*.[3] Their work was not only inspired by weather forecasters, says Friedman, in an interview with *Significance*. "It developed because of the sharp ups and downs of prices in the late nineteenth and early twentieth century, coupled with the desire of businesspeople to make future plans. Babson developed his service, for instance, after the 1907 panic. The way he shaped his forecasting method was reliant on traditions of barometers, ideas about the business cycle, and the availability of data."

Whereas Babson's method looked at trends in data over time, his contemporary, Irving Fisher, built a machine to model how the flow of supply and demand of one commodity affected that of others. It was a hydraulic computer. When Fisher adjusted a lever, water flowed through a series of tubes to restore equilibrium between the prices of goods. A similar device was built by William Phillips at the London School of Economics in 1949. It used channels of colored water to replicate taxes, exports, and spending in the British economy.

Babson was alone among his contemporaries in predicting the 1929 stock market crash. However, says Friedman, he and other early forecasters made assumptions that were simplistic and often misguided.

Wrong though they may have been, the methods used by Babson and Fisher were forerunners of the range of statistical tools available today. Babson worked with time series, plotting the aggregate of variables such as crop production, commodity prices, and business failures on a single chart that forecast how the economy would fare. Fisher's hydrostatic machine, on the other hand, did not include a time element. In Friedman's book, the machine is described as the grandparent of the economic forecasting models developed after World War II and run on computers.

The Science of Forecasting

Techniques that we know today were refined in a variety of different fields, says Rob Hyndman, professor of statistics at Monash University in Australia. "In the last 50 years or so, people started doing more time series models that try to relate the past to the future. They became extremely useful in sales forecasting and in predicting demand for items. In other fields such as engineering, they started trying to build models to predict things like river flow based on rainfall. Models for electricity demand were developed so that they could plan generation capacity."

The science of forecasting got going properly in the 1980s, says Hyndman. "People realised that if you took all of the ideas that people had developed in different fields, and you thought of it as a collection of techniques and overlaid that with analytical and scientific thinking, then forecasting itself could be considered a scientific discipline."

But how do we know whether or not quantitative forecasting will work in any given area? Hyndman, who is editor-in-chief of the *International Journal of Forecasting*, believes that the predictability of an event boils down to three factors. "The first is whether you have an idea of what's driving it—the causal factors. So you might just have data on the particular thing you're interested in, but you don't understand why the fluctuations occur or why the patterns exist in the data. You can still forecast it, but not so well. If you haven't understood the way in which the thing you're interested in reacts to the driving factors, you're going to lose that coupling over time," he says.

The two other factors involved are the availability of data and whether a forecast will itself affect what it is that you are trying to predict. If an exchange rate rise is predicted, for instance, it will affect prices in the real markets, which will end up influencing the rate itself.

These factors aside, a successful forecaster also needs a toolbox of statistical methods and the know-how to pick the right method for a particular situation, says Hyndman (see "Forecasting Techniques Explained"). Statistical models, though, only work in the short term. "They're not very good for long-term forecasting because the big assumption—that the future looks similar to the past—slowly breaks down the further you get into the future," he says.

Forecasting Techniques Explained

Rob Hyndman, professor of statistics at Monash University, outlines the different kinds of statistical forecasting methods

On Time-Series Forecasts

A purely time-series approach just looks at the history of the variable you are interested in and builds a model that describes how that has changed over time. You might look at the history of monthly sales for a company. You look at the trends and seasonality and extrapolate it forward, but you do not take any other information into account.

On Explanatory Models

This is where you relate the thing you are trying to forecast with other things that might affect it. So if you are forecasting sales, you might also take into account population and the state of the economy. All you actually need is for the drivers to be good predictors of the outcome, whether or not there is a direct causal relationship or something more complicated. You can also have models that combine the two. So you have some time series and some other information, and you build a model that puts it all together.

On Probabilistic Forecasts

For a long time, people have been producing predictions with prediction intervals—giving a statement of uncertainty using probabilities. The new development is not just giving an interval but also giving the entire probability distribution as your forecast. So you will say, "Here's a number: the chance of being below this is 1%. Here's another number: the chance of being below this is 2%," and so on over 100 percentiles. But if you are giving an entire probability distribution, you cannot measure the accuracy of your forecast in a simple way. New techniques for measuring forecast accuracy that take this into account have become popular in the past five years or so: these are called probability scoring methods.

"Then there are problems where there's just not really enough data to be able to build good models, or situations that are not reflected at all historically, such as technological changes," says Hyndman. "There's no data available that will tell you what's going to happen."

Superforecasters

Unforeseen developments—whether technological, political, or social—pose an interesting dilemma for those whose job is to anticipate such things: national security agencies, for instance. What if such developments are predictable, not from a single data set or time series, perhaps, but from the aggregated opinions of groups of individuals? In 2011, the Aggregative Contingent Estimation (ACE) program set out to answer that question, with funding from the Intelligence Advanced Research Projects Activity (IARPA). ACE announced a forecasting tournament that would run from September 2011 to June 2015, in which five teams would compete by answering 500 questions on world affairs.

One of the teams was the Good Judgment Project (GJP), which was created by Barbara Mellers and political scientist Philip Tetlock, the man behind the 2006 research on expert predictions. The GJP attracted more than 20,000 wannabe forecasters in its first year. In his 2015 book, *Superforecasting*, Tetlock recalls how the forecasters were asked to predict "if protests in Russia would spread, the price of gold would plummet, the Nikkei would close above 9500, or war would erupt on the Korean peninsula."[4]

The GJP won the tournament hands down. According to Tetlock, it beat a control group by 60% in its first year and 78% in its second year. After that, IARPA decided to drop the others, leaving the GJP as the last team standing.

As described in *Superforecasting*, the GJP continually assigned its participants an ever-changing rating called a "Brier score," which measures the accuracy of predictions on a scale from 0 to 2; the lower the number, the more accurate the prediction. By doing so, they were able to identify the best among them, whom they called superforecasters. Some superforecasters were plucked from the crowd and placed in 12-person "superteams" that could share information with each other. Would the superteams perform any better?

It turns out that they could. Superteams were pitched against teams of regular forecasters. They also competed against individual forecasters, whose forecasts were aggregated to produce an unweighted average prediction. This average represented the "wisdom of the crowds," an idea put forward in 1907 by the statistician Sir Francis Galton, who proposed that the accumulated knowledge of a group of people could be more accurate than individual predictions.

The superteams faced one final group of competitors: forecasters who had been assigned to work as traders in prediction markets, a popular form of forecasting in which people place bets on the outcome they think is most likely to happen. Writing in *Superforecasting*, Tetlock says:

"The results were clear-cut. . . . Teams of ordinary forecasters beat the wisdom of the crowd by about 10%. Prediction markets beat ordinary teams by about 20%. And superteams beat prediction markets by 15% to 30%."[4]

Want to Bet?

The fact that Tetlock's superteams were able to beat the markets was something of a surprise. Futures exchanges, such as the Iowa Electronic Markets (IEM) and PredictIt, have an enviable track record in predicting outcomes, especially political outcomes. In November 2015, Professor Leighton Vaughan Williams, director of the Betting Research Unit at Nottingham Business School, and his co-author Dr James Reade, published "Forecasting Elections," a study that compared prediction markets to more traditional methods.[5] "We got huge amounts of data from Intrade and Betfair, plus statistical modelling, expert opinion and every opinion poll. We compared them over many years and literally hundreds of different elections using state-of-the-art econometrics," says Vaughan Williams. "We found that prediction markets significantly outperformed the other methodologies included in our study, and even more so as you get closer to the event."

Similarly, a study conducted by Joyce E. Berg and colleagues, published in 2008, compared IEM predictions to the results of 964 polls over five U.S. presidential elections since 1988. They found that "the market is closer to the eventual outcome 74% of the time" and that "the market significantly outperforms the polls in every election when forecasting more than 100 days in advance."[6]

So why were prediction markets beaten by superforecasters in Tetlock's research? One reason could be, as Tetlock himself writes, that the prediction markets in his contest lacked "liquidity"—in other words, they did not feature substantial amounts of money or activity. Vaughan Williams believes that liquid markets would normally win for particular kinds of event. "If you ask [superforecasters] to beat a market on who's going to win the Oscars or the Super Bowl or Florida in the 2016 U.S. election they'd find it tough—because millions of pounds will be traded," he says. "But if you ask them: 'Will David Cameron have a meeting with Jean-Claude Juncker by Thursday night?' in that situation there's no real market for them to beat."

Markets are not very good at predicting things that are inherently unpredictable, however. "A prediction market can't aggregate information on something that people can't work out, like the Lottery or earthquakes," says Vaughan Williams. "You know where earthquakes are more likely to happen, but you can't predict on what date they'll occur."

Foreseeing terrorist attacks is something else that markets are just not built for, he adds. "Terrorists are hardly going to be putting their money in and tipping their hand—if anything, they would be doing the opposite."

But where they can be used, prediction markets have another advantage, says Vaughan Williams. "Often what's just as important as knowing what the future will be is knowing it before somebody else. I've just accepted a paper for the *Journal of Prediction Markets* showing that the IEM's influenza market is effective for predicting outbreaks. It's because the market aggregates information from everyone on the system in real time, second by second. If you see everyone around you sneezing with what looks like flu, you could go to your computer and start trading."

Managing Complexity

Predicting epidemics is an area where statistical forecasting methods struggle, says Rob Hyndman. "We just don't have the data on which viruses are brewing and which mosquito populations are breeding, so it's extremely difficult—if not impossible," he says. Google infamously sought a way around this problem by analyzing search activity to predict the spread of the influenza virus. Early estimates were reliable and accurate, but in time the model produced overestimates—in part because it failed to fully account for the fact that flu-related searches might be made by healthy individuals.[7]

One area that has seen steady progress, however, is weather and climate prediction. A review published in *Nature* in September 2015 revealed the "quiet revolution" in numerical weather prediction that has made today's six-day forecasts as accurate as the five-day forecasts of 10 years ago.[8] Forecasters are able to give one to two weeks' lead time for extreme events like Russia's recent heat wave and the U.S. cold snap, while fluctuations in sea surface temperatures following El Niño can be predicted three to four months beforehand.

The improvements have been driven by number-crunching power from supercomputers, backed by a hierarchy of models of varying complexity, and global data from satellites.

However, there are inherent limitations to modeling complex systems like the climate, according to David Orrell, who runs the scientific consultancy Systems Forecasting. "If you look at the formation of a cloud, you have an interaction between minute particles of something that forms a seed for a droplet. The droplet grows, and that process is incredibly non-linear and very sensitive to small changes. It involves things over all scales, from the microscopic scale to the scale of a cloud."

Orrell, whose work involves forecasting the effects of cancer drugs, says that organic systems like the climate and the human body are fundamentally different from the kinds of mechanistic systems we are good at modeling. "The dream is that if we just add more and more levels of detail we'd be able to capture this [behavior], but there's a fundamental limitation to what you can do with mechanistic models."

Whatever Next?

So how will forecasting evolve in the future? In terms of opinion-based predictions, look no further than Almanis, a cross between a prediction market and the Good Judgment Project. Describing itself as a "crowd wisdom platform," Almanis incentivizes forecasters with points, not pounds, although it awards real money prizes to the most accurate users. It is a commercial entity, making money by charging companies or governments to post questions.

Services like Almanis will be commonly used within the next 10 years, believes Leighton Vaughan Williams. "As academic research further improves their efficiency I think they'll become a key part of corporate forecasting and information aggregation. Say I want to reduce the waiting list in an eye clinic and I've got a budget of £100,000. Should I hire a doctor, or two nurses? A prediction market can give you that sort of information."

In terms of mathematical forecasting, Rob Hyndman says that methods are being developed to cope with massive data. "One trend that's happening at the moment is that a lot of the techniques that computer scientists have developed in machine learning in other fields are coming into forecasting. It's very interesting to see what's going on."

Statistical models are also being used to combine other types of predictions into metaforecasts. One example is PredictWise, an academic project started by David Rothschild, an economist at Microsoft Research. The tool combines information from prediction markets, opinion polls, and bookmakers' odds to come up with probabilities for everything from the next James Bond to the likelihood of the U.K. leaving the European Union.

Probabilities are not the same as certainties, however. "In 2008, Hillary Clinton had a 20% chance of winning the New Hampshire primary and she won it," recalls Vaughan Williams. "People said the prediction markets had got it all wrong. But as any statistician would know, what they're saying is that one time in five it's going to happen."

As for mathematical models, Rob Hyndman makes the point that they are always just simplifications of reality—and life is sometimes too complex to model, whether accurately or approximately. The future, or parts of it, therefore, will remain unforeseen. But it is safe to predict that forecasters will keep trying to catch a glimpse of what lies ahead.

References

1. Tetlock, P. E. (2006) *Expert Political Judgment: How Good Is It? How Can We Know?* Princeton, NJ: Princeton University Press.
2. Orrell, D. (2007) *The Future of Everything: The Science of Prediction*. New York: Basic Books.
3. Friedman, W. A. (2014) *Fortune Tellers: The Story of America's First Economic Forecasters*. Princeton, NJ: Princeton University Press.
4. Tetlock, P. E., and Gardner, D. (2015) *Superforecasting: The Art and Science of Prediction*. New York: Crown.
5. Vaughan Williams, L., and Reade, J. (2015) "Forecasting elections." *Journal of Forecasting*. doi: 10.1002/for.2377.
6. Berg, J. E., Nelson, F. D., and Rietz, J. A. (2008) "Prediction market accuracy in the long run." *International Journal of Forecasting*, **24**(2), 285–300.
7. Lazer, D., and Kennedy, R. (2015) "What we can learn from the epic failure of Google Flu Trends." *Wired*, 1 October. bit.ly/1L8xCUK.
8. Bauer, P., Thorpe, A., and Brunet, G. (2015) "The quiet revolution of numerical weather prediction." *Nature*, **525**, 47–55.

Contributors

Gerald L. Alexanderson is on the faculty of Santa Clara University. When a graduate student at Stanford, he became an assistant to George Polya, which turned into a long relationship. Polya did not know Ramanujan, though they were born in the same year, 1887, but Polya did not go to Cambridge until four years after Ramanujan's death in India. Both enjoyed a close relationship with Hardy and Littlewood. Alexanderson served as secretary and president of the Mathematical Association of America (MAA) and is hard at work on a 19th book manuscript; his most recent was a book on G. H. Hardy, published by the MAA and Cambridge in 2016.

Robert Bain is a freelance journalist based in the United Kingdom. After graduating in modern European languages from the University of Edinburgh, he spent five years on the staff of *Research*, the magazine of the U.K.'s Market Research Society, where he was introduced to writing about statistics. He went on to edit the lighting magazines *Lux* and *Lux Review* and has been working freelance since 2015. He now writes for publications including the statistics magazine *Significance*, which is published jointly by the American Statistical Association and the U.K.'s Royal Statistical Society. Recently, Bain has changed his name to Robert Langkjær-Bain.

Viktor Blåsjö is a historian of mathematics at the Mathematical Institute of Utrecht University. His dissertation formed the basis for his monograph *Transcendental Curves in the Leibnizian Calculus* (Elsevier 2017). He is also interested in using a historical perspective to inform mathematics teaching; his website, intellectualmathematics.com, contains his calculus textbook and other materials drawing on this synergy. Follow him on Twitter @viktorblasjo.

Jo Boaler is a professor of mathematics education at Stanford University, and the co-founder of youcubed. She is the author of the first massive online open course (MOOC) on mathematics teaching and learning. Former roles have included being the Marie Curie Professor of Mathematics Education in England, a mathematics teacher in London, and a researcher at King's College, London. Her Ph.D. thesis won the national award for educational research in the United Kingdom. She is a fellow of the Royal Society of Arts, the recipient of an NSF Early Career Award, an NCSM Equity Award, and a CMC Mathematics Leadership award. She is the author of nine books, two TEDx talks, and numerous research articles. She was recently named one of the eight educators "changing the face of education" by the BBC.

Sinead Breen holds a Ph.D. in mathematics (on asymptotic analysis) from Dublin City University and has recently returned there as a mathematics lecturer. She conducts research in mathematics education; her main interest is in the teaching and learning of mathematics at undergraduate level.

Lang Chen, Ph.D., is a research scholar at Stanford University in the Department of Psychiatry and Behavioral Sciences. His research focuses on understanding the development of human brain networks that support memory and learning to establish representations of knowledge. Combining both computational modeling and neuroimaging approaches, he currently works on projects that investigate typical and atypical development of math, language, and social concepts in children. His work has been published in peer-reviewed journals including *Nature Human Behaviour*, *Journal of Cognitive Neuroscience*, and *Journal of Neuroscience*.

Philip J. Davis is known for his books in numerical analysis, computational mathematics, and approximation theory. He has also explored questions in the history and philosophy of mathematics, as well as the role it plays in society. Davis is currently professor emeritus in the Division of Applied Mathematics of Brown University.

Marc Frantz received a B.F.A. in painting from the Herron School of Art in 1975, followed by a 13-year career with art galleries. He earned an M.S. in mathematics from IUPUI in 1990 and is currently a research associate in mathematics at Indiana University. He is coauthor with Annalisa Crannell of the textbook *Viewpoints: Mathematical Perspective and Fractal Geometry in Art* (Princeton University Press 2011).

Jeremy Gray is a professor emeritus at the Open University in the United Kingdom. He is the author of *Plato's Ghost: Modernism and Mathematics* (2008) and *Henri Poincaré: A Scientific Biography* (2013), both published by Princeton University Press. He is a fellow of the American Mathematical Society and has been a recipient of the Leon Whiteman award of the AMS for his work in the history of mathematics. In 2016, he won the Neugebauer Prize of the European Mathematical Society for his historical work.

Kevin Hartnett is the senior math writer for *Quanta Magazine*. From 2013–2016, he wrote "Brainiac," a weekly column on new research for the *Boston Globe*'s Ideas section. A native of Maine, Kevin lives in South Carolina with his wife and three sons. Follow him on Twitter @kshartnett.

Mohammadhossein Kasraei is an architect and researcher in the fields of design and architecture at FabLab Tehran. He received his undergraduate degree in 2010 and his master of architecture in 2012 from the Art University of Tehran before working as a professional architect. He has managed and designed a full range of architectural projects, spanning from interior design to landscape and from initial studies to detailing. He has interdisciplinary research interests, which are demonstrated in his design and research. His research interests are in the areas of Islamic architecture, computational design, and fabrication. He founded 5O7Studio in Tehran in 2011 as an architectural design studio and has been working there as an architect and designer. He also founded FabLab Tehran in 2014, a Tehran-based digital fabrication laboratory, in order to provide an open-access environment for students and designers.

Evelyn Lamb is a freelance math and science writer based in Salt Lake City, Utah. She got her Ph.D. in mathematics in 2012 from Rice University and started doing science writing thanks to a AAAS-AMS mass media fellowship at *Scientific American*. She taught at the University of Utah before leaving academia to devote herself to writing full time. She is a coauthor of the AMS *Blog on Math Blogs* and has written for a variety of publications, including *Scientific American*, *Slate*, *Nature News*, *Smithsonian*, and *Undark*. Her blog, Roots of Unity, is on the *Scientific American* blog network.

Mohammadjavad Mahdavinejad graduated from SAMPAD—National Organization for Development of Exceptional Talents of Iran—in 1996 and then enrolled in the University of Tehran's Department of Architecture. He graduated as a gold medalist of the school in 2001 and continued his Ph.D. at the University of Tehran until graduation as the top student in 2007. He experienced a wide range of postdoctoral work, as well as vocational courses in digitalism, sustainable architecture, and computational energy while participating in international laboratories and workshops. He won some outstanding national and international prizes in academic as well as architectural design competitions. He was recognized as the most cited scholar in the field of architecture and planning in 2009, 2010, 2012, and 2014 by the Ministry of Science, Research and Technology. In 2007, he joined the Department of Architecture at Tarbiat Modares University as academic staff. In 2013, as an associate professor of the Department of Architecture, he started interdisciplinary studies between TMU and other universities in Europe, Asia, and North America. He has published more than 100 scientific papers and 20 books as author, translator, and editor. He has focused his research on scientific and technical studies and high-performance architecture. He has been the dean of the High-Performance Architecture Lab at TMU since 2015.

Jean-Pierre Marquis is currently vice dean of the Faculty of Arts and Sciences at the University of Montreal. He usually is a professor in the Department of Philosophy, where he teaches logic, philosophy of science, and epistemology. His area of research is primarily philosophy and foundations of mathematics. He fell in the cauldron of categories, functors, and natural transformations as an undergraduate while studying at McGill University and never recovered. His publications include a book entitled *From a Geometrical Point of View. A Study in the History and Philosophy of Category Theory*, published in 2009 by Springer. He has published numerous articles on category theory, categorical logic, homotopy theory, and algebraic topology in various venues.

John Mason is a professor emeritus of mathematics education at the Open University, where he worked for 40 years, and Senior Research Fellow at the Department of Education, University of Oxford. He is the principal author of *Thinking Mathematically* (1982), which was predicted to be "a classic" when it first appeared and which is still being used 35 years later; the author of *Researching Your Own Practice: The Discipline of Noticing* (2004); and coauthor of *Mathematics as a Constructive Activity: Learners Generating Examples* (2005), among numerous other books and papers. His focus is on promoting and supporting mathematical thinking and supporting others with similar concerns. His particular interest is the role and nature of attention.

Yahya Nourian is an architect and researcher in the fields of art, design, and architecture. He is graduated from Tehran University of Art, Department of Architecture in 2011 and earned his master of architecture in 2013 from Tarbiat Modares University as a top student in 2013. He has managed and designed a full range of architectural projects, spanning from interior design to landscape and from initial studies to detailing. He has interdisciplinary research interests, which are demonstrated in his design and research. He has been the director of 5O7Studio since he founded it in 2011 and CEO of FabLab Tehran since he founded it in 2014. His research interests are in the areas of history, history of art, Islamic architecture, design, and fabrication, and he has published scientific papers and online articles as author and editor or editor in chief. He has focused his research on studies of the history of art at the Art-story Research Institute, which he founded. He has been the dean of the school since 2013. (E-mail: yahya.nourian@gmail.com)

Ann O'Shea is a senior lecturer in the Department of Mathematics and Statistics at the Maynooth University in Ireland. She received her Ph.D. from the University of Notre Dame, Indiana, in 1991. She began her career by working in the area of value distribution theory in several complex variables. Currently her research interests lie in mathematics education, especially at the undergraduate level. She received a National Award for Teaching Excellence in 2010.

Larry Riddle is a professor of mathematics at Agnes Scott College, a private liberal arts college for women in Decatur, Georgia. He has served as chief reader for the Advanced Placement Calculus exam and also chaired the AP Calculus Development Committee. He is the developer of the website *Biographies of Women Mathematicians*, which illustrates the numerous achievements of women in mathematics, and has created a website on classic iterated function systems and their fractals. Many of his digital fractal prints and cross-stitch fractal artwork have been displayed in the annual Exhibition of Mathematical Art at the Joint Mathematics Meetings.

Siobhan Roberts is a journalist and author whose work focuses on mathematics and science. Currently, she writes for *The New Yorker*.com science page, "Elements," *Quanta,* and *Nautilus.* Her latest book is *Genius at Play: The Curious Mind of John Horton Conway* (Bloomsbury 2015), which won the 2017 JPBM Communications Award for Expository and Popular Books. Her previous books are *Wind Wizard: Alan G. Davenport and the Art of Wind Engineering* (Princeton University Press 2012), and *King of Infinite Space: Donald Coxeter, the Man Who Saved Geometry* (Bloomsbury 2006), which won the Mathematical Association of America's 2009 Euler Prize for expanding the public's view of mathematics.

Carlo H. Séquin has been on the faculty in the EECS Department at UC Berkeley since 1977. He has been in love with geometry since high school and has typically focused on the geometry component in most of his professional assignments. Those tasks included the development of the first TV-compatible solid-state image sensor at Bell Labs, the RISC microcomputer built at UC Berkeley, and the design effort for Soda Hall—the current home of the Computer Science Division. For two decades, he has been teaching courses in computer graphics and in computer-aided design and modeling. More recently, Séquin has used 3D printing to create mathematical visualization models and maquettes for abstract geometrical sculptures.

Raymond Shiau is a software engineer in the San Francisco Bay area. He currently aspires to improve recommendations at Pinterest through the application of machine learning techniques. He holds a B.S. in electrical engineering and computer science from the University of California, Berkeley, and had the privilege of contributing to research during that time, thanks to the Undergraduate Research Apprentice Program (URAP). He is a recipient of the Jim and Donna Gray Endowment Award. Professionally, his interests lie in the fields of machine learning and computer graphics, but in his free time, he enjoys skiing, singing, and long-distance running.

Graham Southorn is a freelance writer and editor specializing in science, technology, and business. He is a former editor of *BBC Focus* magazine, which was twice named magazine of the year at the Digital Magazine Awards. In 2005, he launched *BBC Sky at Night* with Sir Patrick Moore, a magazine he edited for six years. He is the coauthor of *Physics Squared*, published by Apple Press, and has written for *Guardian* blogs, *Significance*, and *HERE 360*, among others. He helped judge the UWE Science Writing Competition in 2016 and was most recently the editor of *South West Business Insider* in Bristol, U.K. He tweets from @GrahamSouthorn

Lloyd Trefethen is a professor of numerical analysis at Oxford University. As an author, he is known for his books, including *Numerical Linear Algebra* (1997), *Spectral Methods in MATLAB* (2000), *Spectra and Pseudospectra* (2005), *Trefethen's Index Cards* (2011), *Approximation Theory and Approximation Practice* (2013), and *Exploring ODEs* (to appear in 2018). Trefethen served as president of the Society for Industrial and Applied Mathematics, SIAM, during 2011–2012 and is a member of the U.S. National Academy of Engineering and a fellow of the Royal Society. He is the creator of the software system Chebfun.

Noson S. Yanofsky has a Ph.D. in mathematics from the Graduate Center of the City University of New York. He is a professor of computer science at Brooklyn College and the Graduate Center. His research is focused on category theory, theoretical computer science, and quantum foundations. He coauthored *Quantum Computing for Computer Scientists* (Cambridge University Press 2008) and is the author of *The Outer Limits of Reason: What Science, Mathematics, and Logic Cannot Tell Us* (MIT Press 2013). Noson lives in Brooklyn with his wife and four children.

Notable Writings

This volume is not just an anthology; it is also a reference source for other notable writings on mathematics published in 2016. The first list below can offer you many more opportunities to explore what people from different walks of life and many disciplines think and write about mathematics. Many of these pieces caught my attention while reaching the final selection for the book.

Several other lists follow: of remarkable book reviews, interviews with mathematical people, memorial notes, and special journal issues. None of these lists is comprehensive; to save space, in some cases I omitted full bibliographic references for materials.

Abadi, Daniel. "The Beckman Report on Database Research." *Communications of the ACM* 59.2(2016): 92–99.

Abbasian, Reza O., and John T. Sieben. "Creating Math Videos: Comparing Platforms and Software." *PRIMUS* 26.2(2016): 168–77.

Adams, Marcus P. "Hobbes on Natural Philosophy as "True Physics" and Mixed Mathematics." *Studies in History and Philosophy of Science* 56(2016): 43–51.

Aguilar, Marion Sánchez, et al. "Exploring High-Achieving Students' Images of Mathematicians." *International Journal of Science and Mathematics Education* 14(2016): 527–48.

Akkoç, Hatice, Mehmet Ali Balkanlıoğlu, and Sibel Yeşildere-İmre. "Exploring Preservice Mathematics Teachers' Perception of the Mathematics Teacher through Communities of Practice." *Mathematics Teacher Education and Development* 18.1(2016): 37–51.

Aksoy, Asuman Güven. "Al-Khwārizmī and the Hermeneutic Circle: Reflections on a Trip to Samarkand." *Journal of Humanistic Mathematics* 6.2(2016): 114–27.

Alexanderson, Gerald L., and Leonard F. Klosinski. "Gauss and the '*Disquisitiones Arithmeticæ*'." *Bulletin of the American Mathematical Society* 53.1(2016): 117–20.

Alexanderson, Gerald L., and Leonard F. Klosinski. "The Newton-Leibniz Controversy." *Bulletin of the American Mathematical Society* 53.2(2016): 295–99.

Amador, Julie. M. "Teachers' Considerations of Students' Thinking during Mathematics Lesson Design." *School Science and Mathematics* 116.5(2016): 239–52.

Amir, Ariel, Mikhail Lemeshko, and Tadashi Tokieda. "Surprises in Numerical Expressions of Physical Constants." *The American Mathematical Monthly* 123.6(2016): 609–12.

Anatriello, Giuseppina, Francesco Saverio Tortoriello, and Giovanni Vincenzi. "On an Assumption of Geometric Foundation of Numbers." *International Journal of Mathematical Education in Science and Technology* 47.3(2016): 395–407.

Apostol, Tom M., and Mamikon A. Mnatsakanian. "A New Look at Surfaces of Constant Curvature." *The American Mathematical Monthly* 123.5(2016): 439–47.

Arcavi, Abraham. "Revisiting Aspects of Visualization in Mathematics Education." *La Matematica nella Società e nella Cultura* 8(2015): 143–60.

Armistead, Timothy W. "Misunderstood and Unattributed: Revisiting M. H. Doolittle's Measures of Association, with a Note on Bayes' Theorem." *The American Statistician* 70.1(2016): 63–73.

Asorey, Manuel. "Space, Matter and Topology." *Nature Physics* 12(2016): 616–18.

Bain, Robert. "Citizen Science and Statistics: Playing a Part." *Significance* 13.1(2016): 17–21.

Baker, Alan. "Parsimony and Inference to the Best Mathematical Explanation." *Synthese* 193(2016): 333–50. [see also Pataut]

Bakhshi, Hasan. "How Can We Measure the Modern Digital Economy?" *Significance* 13.3(2016): 6–7.

Baquero, Carlos, and Nuno Preguiça. "Why Logical Clocks Are Easy." *Communications of the ACM* 59.4(2016): 43–47.

Baranyi, Michael J. "Fellow Travelers and Traveling Fellows: The Intercontinental Shaping of Modern Mathematics in Mid-Twentieth Century Latin America." *Historical Studies in the Natural Sciences* 46(2016): 669–709.

Barnett, Janet Heine, Jerry Lodder, and David Pengelley. "Teaching and Learning Mathematics from Primary Historical Sources." *PRIMUS* 26.1(2016): 1–18.

Baron, Sam. "Mathematical Explanation and Epistemology: Please Mind the Gap." *Ratio* 29(2016): 149–66.

Baron, Sam. "The Explanatory Dispensability of Idealizations." *Synthese* 193(2016): 365–86. [see also De Bianchi]

Barrow-Green, June, and Reinhard Siegmund-Schultze. "'The First Man on the Street'—Tracing a Famous Hilbert Quote (1900) to Gergonne (1825)." *Historia Mathematica* 43(2016): 415–26.

Barwell, Richard. "Formal and Informal Mathematical Discourses: Bakhtin and Vygotsky, Dialogue and Dialectic." *Educational Studies in Mathematics* 92(2016): 331–45.

belcastro, sarah-marie. "Do the Twist! (on Polygon-Base Boxes)." *The College Mathematics Journal* 47.5(2016): 340–45.

Bentley, Brendan. "Why the Golden Proportion Is Really Golden." *Australian Mathematics Teacher* 72.1(2016): 10–14.

Benvenuti, Silvia, and Linda Pagli. "Refrigerator Ladies." *Matematica, Cultura e Società* 1.1(2016): 51–64.

Beveridge, Andrew, and Jie Shan. "Network of Thrones." *Math Horizons* 23.4(2016): 18–22.

Bishop, Jessica Pierson, et al. "Leveraging Structure: Logical Necessity in the Context of Integer Arithmetic, Mathematical Thinking and Learning." *Mathematical Thinking and Learning* 18.3(2016): 209–32.

Blåsjö, Viktor. "In Defense of Geometrical Algebra." *Archive for History of Exact Sciences* 70.3(2016): 325–59.

Blåsjö, Viktor. "The How and Why of Constructions in Classical Geometry." *Nieuw Archief voor Wiskunde* 5/17.3(2016): 283–91.

Boaler, Jo. "Why Math Education in the U.S. Doesn't Add Up." *Scientific American Mind* 27.6(2016): 18–19.

Bonomo, John P. "Winning a Pool Is Harder Than You Thought." *The College Mathematics Journal* 47.5(2016): 347–55.

Borchert, Carol Ann, and Jason Boczar. "*Numeracy* and Evaluating Quality in Open Access Journals." *Numeracy* 9.2(2016).

Borrell, Brendan. "Physics on Two Wheels." *Nature* 535(2016): 338–41.

Bossé, Michael, Eric Marland, Gregory Rhoads, and Michael Rudziewicz. "Searching for the Black Box: Misconceptions of Linearity." *Chance* 29.4(2016): 14–23.

Boumans, Marcel. "Graph-Based Inductive Reasoning." *Studies in History and Philosophy of Science, A* 59(2016): 1–10.

Boylan, Mark. "Ethical Dimensions of Mathematics Education." *Educational Studies in Mathematics* 92(2016): 395–409.

Bozzo, Enrico, and Massimo Franceschet. "A Theory of Power in Networks." *Communications of the ACM* 59.11(2016): 75–83.

Brandt, Jim, Jana Lunt, and Gretchen Rimmasch Meilstrup. "Mathematicians' and Math Educators' Views on 'Doing Mathematics.'" *PRIMUS* 26.8(2016): 753–69.

Breitenbach, Angela. "Beauty in Proofs: Kant on Aesthetics in Mathematics." *European Journal of Philosophy* 23.4(2016): 955–77.

Brown, Tony. "Rationality and Belief in Learning Mathematics." *Educational Studies in Mathematics* 92(2016): 75–90.

Burn, Bob. "Early Tables Resembling Those of Natural Logarithms." *BSHM Bulletin—Journal of the British Society for the History of Mathematics* 31.2(2016): 112–22.

Bussotti, Paolo. "La concezione dell'infinito in Federigo Enriques." *Matematica, Cultura e Società* 1.1(2016): 65–86.

Butler, Steve, Persi Diaconis, and Ron Graham. "The Mathematics of the Flip and Horseshoe Shuffles." *The American Mathematical Monthly* 123.6(2016): 542–56.

Buvinic, Mayra, and Ruth Levine. "Closing the Gender Data Gap." *Significance* 13.2(2016): 34–37.

Caglayan, Günhan. "Dramathizing Functions: Building Connections between Mathematics and Arts." *Journal of Humanistic Mathematics* 6.1(2016): 235–41.

Caglayan, Günhan. "Exploring the Lunes of Hippocrates in a Dynamic Geometry Environment." *BSHM Bulletin—Journal of the British Society for the History of Mathematics* 31.2(2016): 144–53.

Calderón-Tena, Carlos O., and Linda C. Caterino. "Mathematics Learning Development: The Role of Long-Term Retrieval." *International Journal of Science and Mathematics Education* 14(2016): 1377–85.

Cartwright, Julyan H. E., and Diego L. González. "Möbius Strips before Möbius: Topological Hints in Ancient Representations." *The Mathematical Intelligencer* 38.2(2016): 69–76.

Castelvecchi, Davide. "The 'Family Trees' of Mathematics." *Nature* 537(2016): 20–21.

Castera, Jean-Marc. "Persian Variations." *Nexus Network Journal* 18(2016): 223–74.

Caulfield, Michael J. "Maine v. Mathematics: A Tale of Reapportionment, Calculation and Indignation." *The Mathematical Intelligencer* 38.2(2016): 59–64.

Chen, Yiling, et al. "Mathematical Foundations for Social Computing." *Communications of the ACM* 59.12(2016): 102–8.

Choy, Bang Heng. "Snapshots of Mathematics Teacher Noticing during Task Design." *Mathematics Education Research Journal* 28(2016): 421–40.

Ciosek, Marianna, and Maria Samborska. "A False Belief about Fractions—What Is Its Source?" *The Journal of Mathematical Behavior* 42(2016): 20–32.

Cirillo, Pasquale, and Nassim Nicholas Taleb. "What Are the Chances of War?" *Significance* 13.6(2016): 44–45.

Clader, Emily. "What If? Mathematics, Creative Writing, and Play." *Journal of Humanistic Mathematics* 6.1(2016): 212–19.
Clark, David. "Seeking *Sangaku*: Visiting Japan's Homegrown Mathematics." *Math Horizons* 24.2(2016): 8–11.
Clark, Kathleen M., and Emmet P. Harrington. "The Paul A. M. Dirac Papers at Florida State University: A Search for Informal Mathematical Investigations." *BSHM Bulletin—Journal of the British Society for the History of Mathematics* 31.3(2016): 205–14.
Clarke, Christopher. "Multi-Level Selection and the Explanatory Value of Mathematical Decompositions." *British Journal for the Philosophy of Science* 67(2016): 1025–55.
Clement, Richard. "Eye Movement Space." *The Mathematical Intelligencer* 38.1(2016): 8–13.
Cochran, Jill A., et al. "Expanding Geometry Understanding with 3-D Geometry." *Mathematics Teaching in the Middle School* 21.9(2016): 534–42.
Conway, John H. "Chemical π." *The Mathematical Intelligencer* 38.4(2016): 7–10.
Conway, John H., and Alex Ryba. "Remembering Spherical Trigonometry." *Mathematical Gazette* 100.547(2016): 1–8.
Corry, Leo. "Some Distributivity-Like Results in the Medieval Arithmetic of Jordanus Nemorarius and Campanus de Novara." *Historia Mathematica* 43(2016): 310–31.
Cox, Teodora. "Integrating Literature in the Teaching of Mathematics." *Australian Mathematics Teacher* 72.1(2016): 15–17.
Crabtree, Jonathan. "Squaring the Circle: A Practical Approach." *Vinculum* 53.4(2016): 7–9.
Crede, Carsten J. "Getting a Fix on Price-Fixing Cartels." *Significance* 13.2(2016): 38–41.
Cretney, Rosanna. "Editing and Reading Early Modern Mathematical Texts in the Digital Age." *Historia Mathematica* 43(2016): 87–97.
Cromwell, Peter R. "Modularity and Hierarchy in Persian Geometric Ornament." *Nexus Network Journal* 18(2016): 7–54.
Damiano, David, Stephan Ramon Garcia, and Michele Intermont. "Liberal Arts Colleges: An Overlooked Opportunity." *Notices of the American Mathematical Society* 63.5(2016): 565–70.
Dasgupta, Shamik. "Symmetry as an Epistemic Notion (Twice Over)." *British Journal for the Philosophy of Science* 67(2016): 837–78.
Date, Sachin. "Should You Upload or Ship Big Data to the Cloud?" *Communications of the ACM* 59(2016): 44–51.
Dawes, Jonathan H. P. "After 1952: The Later Development of Alan Turing's Ideas on the Mathematics of Pattern Formation." *Historia Mathematica* 43(2016): 49–64.
Dawkins, Paul Christian, and Shiv Smith Karunakaran. "Why Research on Proof-Oriented Mathematical Behavior Should Attend to the Role of Particular Mathematical Context." *The Journal of Mathematical Behavior* 44(2016): 65–75.
Dawson, Ryan. "Wittgenstein on Set Theory and the Enormously Big." *Philosophical Investigations* 39(2016): 313–34.
De Bianchi, Silvia. "Which Explanatory Role for Mathematics in Scientific Models?" *Synthese* 193(2016): 387–401. [see also Baron]
De Frejtas, Elizabeth. "Material Encounters and Media Events: What Kind of Mathematics Can a Body Do?" *Educational Studies in Mathematics* 91(2016): 185–202.
De Risi, Vincenzo. "The Development of Euclidean Axiomatics." *Archive for History of Exact Sciences* 70.5(2016): 591–676.
Del Centina, Andrea. "On Kepler's system of conics in 'Astronomiae pars optica.'" *Archive for History of Exact Sciences* 70.1(2016): 567–89.
Del Centina, Andrea. "Poncelet's Porism: A Long Story of Renewed Discoveries." *Archive for History of Exact Sciences* 70.1(2016): 1–122 and 70.2(2016): 123–73.

Despeaux, Sloan Evans, and Adrian C. Rice. "Augustus De Morgan's Anonymous Reviews for 'The Athenæum': A Mirror of a Victorian Mathematician." *Historia Mathematica* 43(2016): 148–71.

Dietiker, Leslie. "The Role of Sequence in the Experience of Mathematical Beauty." *Journal of Humanistic Mathematics* 6.1(2016): 152–73.

Ding, Meixia. "Opportunities to Learn: Inverse Relations in U.S. and Chinese Textbooks, Mathematical Thinking and Learning." *Mathematical Thinking and Learning* 18.1(2016): 45–68.

Duchin, Moon. "Counting in Groups: Fine Asymptotic Geometry." *Notices of the American Mathematical Society* 63.8(2016): 871–74.

Dunham, William. "Bertrand Russell at Bryn Mawr." *The Mathematical Intelligencer* 38.3(2016): 30–40.

Dworkin, Myles, and Elyn Rykken. "Constructions on the Sphere." *Math Horizons* 23.4(2016): 8–11.

Eder, Günter. "Frege's 'On the Foundations of Geometry' and Axiomatic Metatheory." *Mind* 125.497(2016): 5–40.

Edwards, Chris. "Reconciling Quantum Physics with Math." *Communications of the ACM* 59.10(2016): 11–13.

Eilers, Søren. "The LEGO Counting Problem." *The American Mathematical Monthly* 123.5(2016): 415–26.

English, Lyn D. "Revealing and Capitalizing on Young Children's Mathematical Potential." *Zentralblatt für Didaktik der Mathematik* 48(2016): 1079–87.

Ennis, Christopher. "(Always) Room for One More." *Math Horizons* 23.3(2016): 8–12.

Erickson, Ander W. "Rethinking the Numerate Citizen: Quantitative Literacy and Public Issues." *Numeracy* 9.2(2016).

Ernest, Paul. "A Dialogue on the Ethics of Mathematics." *The Mathematical Intelligencer* 38.3(2016): 69–77.

Ernest, Paul. "The Problem of Certainty in Mathematics." *Educational Studies in Mathematics* 92(2016): 379–93.

Ernest, Paul. "The Unit of Analysis in Mathematics Education: Bridging the Political-Technical Divide?" *Educational Studies in Mathematics* 92(2016): 37–58.

Esquincalha, Agnaldo Da C., and Celina A. A. P. Abar. "Knowledge Revealed by Tutors in Discussion Forums with Math Teachers." *Teaching Mathematics and Its Applications* 35(2016): 65–73.

Fenyvesi, Kristóf. "Bridges: A World Community for Mathematical Art." *The Mathematical Intelligencer* 38.2(2016): 35–45.

Folkerts, Menso, Dieter Lunert, and Andreas Thom. "Jost Bürgi's Method for Calculating Sines." *Historia Mathematica* 43(2016): 133–47.

Franchella, Miriam. "In the Footsteps of Julius König's Paradox." *Historia Mathematica* 43(2016): 65–86.

Franklin, James. "Logical Probability and the Strength of Mathematical Conjecture." *The Mathematical Intelligencer* 38.3(2016): 14–19.

Fraser, James D. "Spontaneous Symmetry Breaking in Finite Systems." *Philosophy of Science* 83.3(2016): 585–605.

Fresco, Nir, and Michaelis Michael. "Information and Veridicality: Information Processing and the Bar-Hillel/Carnap Paradox." *Philosophy of Science* 83.1(2016): 131–51.

Friedman, Michael. "Two Beginnings of Geometry and Folding: Hermann Wiener and Sundara Row." *BSHM Bulletin: Journal of the British Society for the History of Mathematics* 31.1(2016): 52–68.

Friend, Michéle, and Daniele Molinini. "Using Mathematics to Explain a Scientific Theory." *Philosophia Mathematica* 24(2016).

Fuller, Joanne. "A New Model of Multiplication via Euclid." *Vinculum* 53.2(2016): 16–18, 21.

Galligan, Linda. "Creating Words in Mathematics." *Australian Mathematics Teacher* 72.1(2016): 20–29.

Gandon, Sébastien. "Rota's Philosophy in Its Mathematical Context." *Philosophia Mathematica* 24(2016).

Garelick, Barry. "Traditional Math: The Exception or the Rule?" *Education News Online* January 4, 2016, http://www.educationnews.org/education-policy-and-politics/traditional-math-the-exception-or-the-rule/.

Garofalo, Vincenza. "The Geometry of a Domed Architecture: A Stately Example of Kārbandi at Bagh-e-Dolat Abad in Yazd." *Nexus Network Journal* 18(2016): 169–95.

Gasparini, Luciano. "La Chiesa del Villaggio del Sole a Vicenza." *Matematica, Cultura e Società* 1.1(2016): 161–71.

George, C. Yousuf, Matt Koetz, and Heather A. Lewis. "Improving Calculus II and III through the Redistribution of Topics." *PRIMUS* 26.3(2016): 241–49.

Giaquinto, Marcus. "Mathematical Proofs: The Beautiful and the Explanatory." *Journal of Humanistic Mathematics* 6.1(2016): 52–72.

Giovannini, Eduardo N. "Bridging the Gap between Analytic and Synthetic Geometry: Hilbert's Axiomatic Approach." *Synthese* 193(2016): 31–70.

Gogate, Vibhav, and Pedro Dominguez. "Probabilistic Theorem Proving." *Communications of the ACM* 59.7(2016): 107–15.

Goldberg, Lisa R. "Mathematical Software: Is It Mathematics or Is It Software?" *Notices of the American Mathematical Society* 63.11(2016): 1293–96.

Goldberg, Timothy E. "Algebra from Geometry in the Card Game SET." *The College Mathematics Journal* 47.4(2016): 265–73.

Goldstone, Robert L., and Gary Lupyan. "Discovering Psychological Principles by Mining Naturally Occurring Data Sets." *Topics in Cognitive Science* 9(2016): 548–68.

Golnabi, Laura. "Creativity and Insight in Problem Solving." *Journal of Mathematics Education at Teachers College* 7.2(2016): 27–29.

Goodman, Noah D., and Michael C. Frank. "Pragmatic Language Interpretation as Probabilistic Inference." *Trends in Cognitive Sciences* 20.11(2016): 818–29.

Goodman, William. "The Promises and Pitfalls of Benford's Law." *Significance* 13.3(2016): 38–41.

Goos, Merrilyn. "Challenges and Opportunities in Teaching Mathematics." *Australian Mathematics Teacher* 72.4(2016): 34–38.

Granberg, Carina. "Discovering and Addressing Errors during Mathematical Problem-Solving—A Productive Struggle?" *The Journal of Mathematical Behavior* 42(2016): 33–48.

Grant, Melva R., William Crombie, Mary Enderson, and Nell Cobb. "Polynomial Calculus: Rethinking the Role of Calculus in High Schools." *International Journal of Mathematical Education in Science and Technology* 47.6(2016): 823–36.

Gray, Mary W. "Statistics Go to School and Then to Court." *Chance* 29.2(2016): 58–60.

Greengard, Samuel. "Graphical Processing Units Reshape Computing." *Communications of the ACM* 59.10(2016): 14–16.

Grima, Pere, Lourdes Roder, and Xavier Tort-Martorell. "Explaining the Importance of Variability to Engineering Students." *The American Statistician* 70.2(2016): 138–142.

Guicciardini, Nicoló. "Lost in Translation? Reading Newton on Inverse-Cube Trajectories." *Archive for History of Exact Sciences* 70.2(2016): 205–41.

Hales, Steven D. "Why Every Theory of Luck Is Wrong." *Nous* 50(2016): 490–508.
Harrison, Steven. "Journalists, Numeracy and Cultural Capital." *Numeracy* 9.2(2016).
Hartimo, Mirja, and Mitsuhiro Okad. "Syntactic Reduction in Husserl's Early Phenomenology of Arithmetic." *Synthese* 193(2016): 937–69.
Hartmann, Heinrich. "Statistics for Engineers." *Communications of the ACM* 59.7(2016): 58–66.
Hartnett, Kevin. "Physicists Uncover Strange Numbers in Particle Collisions." *Wired Online* November 20, 2016, https://www.wired.com/2016/11/physicists-uncover-strange-numbers-particle-collisions/.
Hefferon, Jim, and Alber Schueller. "Writing an Open Text." *The Mathematical Intelligencer* 38.2(2016): 7–9.
Hemenway, Brett. "A New Era for Data Sharing?" *Significance* 13.3(2016): 8–9.
Herbel-Eisenmann, Beth, et al. "Positioning Mathematics Education Researchers to Influence Storylines." *Journal for Research in Mathematics Education* 47.2(2016): 102–17.
Hertel, Joshua T. "Investigating the Implementing Mathematics Curriculum of New England Navigation Cyphering Books." *For the Learning of Mathematics* 36.3(2016): 4–10.
Hipolito, Inês. "Plato on the Foundations of Modern Theorem Provers." *Mathematics Enthusiast* 13.3(2016): 303–15.
Holdener, J., and R. Milnike. "Group Activities for Math Enthusiasts." *PRIMUS* 26.9(2016): 848–62.
Hollings, Christopher. "A Tale of Mathematical Myth-Making: E. T. Bell and the 'Arithmetization of Algebra.'" *BSHM Bulletin: Journal of the British Society for the History of Mathematics* 31.1(2016): 69–80.
Holm, Tara, and Karen Saxe. "A Common Vision for Undergraduate Mathematics." *Notices of the American Mathematical Society* 63.6(2016): 630–34. [also see Jorgensen]
Honner, Patrick. "I Love Teaching Math: Maybe You Will Too." *Math Horizons* 24.2(2016): 34.
Howson, Geoffrey. "Some Thoughts on Educating More Able Students." *Journal of Mathematics Education at Teachers College* 7.1(2016): 1–5.
Huylebrouck, Dirk. "Drifting Runway Numbers." *The Mathematical Intelligencer* 38.5(2016): 52–55.
Huylebrouck, Dirk. "The Diabolic Connection between Paris and Brussels." *The Mathematical Intelligencer* 38.2(2016): 77–79.
Isaacs, Yoaav. "Probabilities Cannot Be Rationally Neglected." *Mind* 125.499(2016): 759–62.
Jaspers, Dany, and Peter A. M. Seuren. "The Square of Opposition in Catholic Hands: A Chapter in the History of 20th Century Logic." *Logique & Analyse* 233(2016): 1–35.
Jooganah, Kamila, and Julian S. Williams. "Contradictions between and within School and University Activity Systems Helping to Explain Students' Difficulty with Advanced Mathematics." *Teaching Mathematics and Its Applications* 35(2016): 159–71.
Jorgensen, Marcus. "Are Mathematics Faculty Ready for Common Vision?" *Notices of the American Mathematical Society* 63.10(2016): 1186–88. [also see Holm and Saxe]
Josefson, Martin. "On the Classification of Convex Quadrilaterals." *Mathematical Gazette* 100.547(2016): 68–85.
Jourdan, Nicholas, and Oleksiy Yevdokimov. "On the Analysis of Indirect Proofs: Contradiction and Contraposition." *Australian Senior Mathematics Journal* 30.1(2016): 55–64.
Kalai, Gil. "The Quantum Computer Puzzle." *Notices of the American Mathematical Society* 63.5(2016): 508–16.
Karaali, Gizem, Edwin H. Villafane Hernandez, and Jeremy A. Taylor. "What's in a Name? A Critical Review of Definitions of Quantitative Literacy, Numeracy, and Quantitative Reasoning." *Numeracy* 9.1(2016).

Kashlak, Adam B. "The Frequency of 'America' in America." *Significance* 13.5(2016): 26–29.

Katz, Eugene A., and Bih-Yaw Jin. "Fullerenes, Polyhedra, and Chinese Guardian Lions." *The Mathematical Intelligencer* 38.3(2016): 61–68.

Kidron, Ivy. "Epistemology and Networking Theories." *Educational Studies in Mathematics* 91(2016): 149–63.

Kiefer, Daniel, Jerome Sacks, and Donald Ylvisaker. "Statistics, Civil Rights and the US Supreme Court: A Cautionary Tale." *Significance* 13.2(2016): 30–33.

Kim, Namjoong. "A Dilemma for the Imprecise Bayesian." *Synthese* 193(2016): 1681–702.

Klarreich, Erica. "Mathematicians Discover Prime Conspiracy." *Quanta Magazine* March 13, 2016, https://www.quantamagazine.org/20160313-mathematicians-discover-prime-conspiracy/.

Klarreich, Erica. "The Oracle of Arithmetic [Peter Scholze]." *Quanta Magazine* June 28, 2016, https://www.quantamagazine.org/20160628-peter-scholze-arithmetic-geometry-profile/.

Koliji, Hooman. "Gazing Geometries: Modes of Design Thinking in Pre-Modern Central Asia and Persian Architecture." *Nexus Network Journal* 18(2016): 105–32.

Kontorovich, Igor', and Boris Koichu. "A Case Study of an Expert Problem Poser for Mathematics Competitions." *International Journal of Science and Mathematics Education* 14(2016): 81–99.

Köse, Emek, and Angela C. Johnson. "Women in Mathematics: A Nested Approach." *PRIMUS* 26.7(2016): 676–93.

Kosko, Karl Wesley. "Writing in Mathematics: A Survey of K–12 Teachers' Reported Frequency in the Classroom." *School Science and Mathematics* 116.5(2016): 276–85.

Koutis, Ioannis, and Ryan Williams. "Algebraic Fingerprints for Faster Algorithms." *Communications of the ACM* 59.1(2016): 98–105.

Kugler, Logan. "What Happens When Big Data Blunders." *Communications of the ACM* 59.6(2016): 15–16.

Kwok, Roberta. "How to Design a Marijuana-License Lottery." *The New Yorker Online* March 22, 2016, http://www.newyorker.com/tech/elements/how-to-design-a-marijuana-license-lottery.

Lacitignola, Deborah. "The Mathematical Beauty of Nature and Turing Pattern Formation." *Matematica, Cultura e Società* 1.1(2016): 93–103.

Lamb, Evelyn. "Math Proof Smashes Size Record." *Nature* 534(2016): 17–18.

Lange, Marc. "Explanatory Proofs and Beautiful Proofs." *Journal of Humanistic Mathematics* 6.1(2016): 9–50.

Lazar, Nicole. "Imagining Genetics: A Tale of Two Modalities." *Chance* 29.2(2016): 61–63.

Lê, François. "Reflections on the Notion of Culture in the History of Mathematics: The Example of 'Geometrical Equations.'" *Science in Context* 29.3(2016): 273–304.

Leeds, Adam E. "Dreams in Cybernetic Fugue: Cold War Technoscience, the Intelligentsia, and the Birth of Soviet Mathematical Economics." *Historical Studies in the Natural Sciences* 46(2016): 633–68.

Lemmermeyer, Franz. "Leonardo da Vinci's Proof of the Pythagorean Theorem." *The College Mathematics Journal* 47.5(2016): 361–64.

Lenharth, Andrew, Donald Nguyen, and Keshav Pigali. "Parallel Graph Analytics." *Communications of the ACM* 59.5(2016): 78–87.

Lew, Kristen, et al. "Lectures in Advanced Mathematics: Why Students Might Not Understand What the Mathematics Professor Is Trying to Convey." *Journal for Research in Mathematics Education* 47.2(2016): 162–98.

Lewis, Nicholas. "Peering through the Curtain: Soviet Computing through the Eyes of Western Experts." *IEEE Annals of the History of Computing* 38.1(2016): 34–47.

Libertini, Jessica, Caitlin Krul, and Erica Turner. "Exam Corrections: A Dual-Purpose Approach." *PRIMUS* 26.9(2016): 803–10.

Livnat, Adi, and Christos Papadimitriou. "Sex as an Algorithm." *Communications of the ACM* 59.11(2016): 84–93.

Llewellyn, Anna. "Problematising the Pursuit of Progress in Mathematics Education." *Educational Studies in Mathematics* 92(2016): 299–314.

Løhre, Erik, and Karl Halvor Teigen. "There Is a 60% Probability, but I Am 70% Certain: Communicative Consequences of External and Internal Expressions of Uncertainty." *Thinking and Reasoning* 23.4(2016): 369–96.

Lombardi, Olimpia, Federico Holik, and Leonardo Vann. "What Is Shannon Information?" *Synthese* 193(2016): 1983–2012.

Lorenat, Jemma. "Synthetic and Analytic Geometries in the Publications of Jakob Steiner and Julius Plücker (1827–1829)." *Archive for History of Exact Sciences* 70.3(2016): 413–62.

Lynn, Henry S. "Training the Next Generation of Statisticians: From Head to Heart." *The American Statistician* 70.2(2016): 149–51.

Lyons, Christopher. "The Secret Life of 1/*n*: A Journey Far beyond the Decimal Point." *Mathematics Enthusiast* 13.3(2016): 189–216.

Maciejewski, Wes, and Bill Barton. "Mathematical Foresight: Thinking in the Future to Work in the Present." *For the Learning of Mathematics* 36.3(2016): 25–30.

Mahboubi, Assia. "Machine-Checked Mathematics." *Nieuw Archief voor Wiskunde* 5/17.3(2016): 172–76.

Maheux, Jean-François. "Wabi-Sabi Mathematics." *Journal of Humanistic Mathematics* 6.1(2016): 174–95.

Malpangotto, Michaela. "The Original Motivation for Copernicus's Research." *Archive for History of Exact Sciences* 70.3(2016): 361–411.

Maltenfort, Michael. "A Canine Conundrum, or What Would Elvis Do?" *The College Mathematics Journal* 47.2(2016): 106–7.

Martin, Lyndon C., and Jo Towers. "Folding Back, Thickening and Mathematical Met-Befores." *The Journal of Mathematical Behavior* 43(2016): 89–97.

Martínez-Planell, Rafael, and Angel Cruz Delgado. "The Unit Circle Approach to the Construction of Sine and Cosine Functions and Their Inverses: An Application of APOS Theory." *The Journal of Mathematical Behavior* 43(2016): 111–33.

Martins, Rogério. "Why Are We Not Able to See Beyond Three Dimensions?" *The Mathematical Intelligencer* 38.4(2016): 46–51.

Maslà, Ramon. "A New Reading of Archytas' Doubling of the Cube and Its Implications." *Archive for History of Exact Sciences* 70.2(2016): 175–204.

Matsudaira, Kate. "The Paradox of Autonomy and Recognition." *Communications of the ACM* 59.3(2016): 55–57.

Mathieu, Romain, et al. "Running the Number Line: Rapid Shifts of Attention in Single-Digit Arithmetic." *Cognition* 146(2016): 229–39.

Matthews, Michael E. "Using IBL in a History of Mathematics Course: A Skeptic's Success." *Journal of Humanistic Mathematics* 6.2(2016): 23–37.

Matthews, Robert. "Beautiful, but Dangerous." *Significance* 13.3(2016): 30–31.

Merrow, Katharine. "Packing Balls in 3, 8, and 24 Dimensions." *Math Horizons*, 24.2(2016): 22–24.

Metcalf, Jacob. "Big Data Analytics and Revision of the Common Rule." *Communications of the ACM* 59.7(2016): 31–33.

Monroe, Christopher R., Robert J. Schoelkopf, and Mikhail D. Lukin. "Quantum Connections." *Scientific American* 314.5(2016): 50–57.

Moskowitz, Carla. "Elegant Equations." *Scientific American* 314.1(2016): 70–73.

Moursund, David. "Joy of Learning and Using Math." *Information Age Education Newsletter* 186(May 2015).

Mozaffari, Mohammad S. "A Forgotten Solar Model." *Archive for History of Exact Sciences* 70.3(2016): 267–91.

Munchack, John Byron. "On Gödel and the Ideality of Time." *Philosophy of Science* 83.4(2016): 1050–58.

Murphy, Carol. "Changing the Way to Teach Maths: Preservice Primary Teachers' Reflections on Using Exploratory Talk in Teaching Mathematics." *Mathematics Teacher Education and Development* 18.2(2016): 29–47.

Murphy, Donal. "George Boole and Walsh's Delusions." *BSHM Bulletin—Journal of the British Society for the History of Mathematics* 31.2(2016): 123–27.

Nahin, Paul J. "The Mysterious Mr. Graham." *The Mathematical Intelligencer* 38.1(2016): 49–55.

Nasar, Audrey A. "The History of Algorithmic Complexity." *Mathematics Enthusiast* 13.3(2016): 217–42.

Neumann, Peter M. "Inspiring Teachers." *Mathematical Gazette* 100.549(2016): 386–95.

Nieder, Andreas. "Representing Something Out of Nothing: The Drawing of Zero." *Trends in Cognitive Sciences* 20.11(2016): 830–42.

Norton, H. James. "Flawed Forensics: Statistical Failings of Microscopic Hair Analysis." *Significance* 13.2(2016): 26–29.

Nye, Mary Jo. "The Republic vs. the Collective: Two Histories of Collaboration and Competition in Modern Science." *NTM Journal of the History of Science, Technology and Medicine* 24(2016): 169–94.

Öhman, Lars-Daniel. "A Beautiful Proof by Induction." *Journal of Humanistic Mathematics* 6.1(2016): 73–85.

Olteanu, Lucian. "Opportunity to Communicate: The Coordination between Focused and Discerned Aspects of the Object of Learning." *The Journal of Mathematical Behavior* 44(2016): 1–12.

Oosterhoff, Richard J. "Lovers in Paratexts: Oronce Fine's Republic of Mathematics." *Nuncius* 31(2016): 549–83.

Orzack, Steven Hecht. "Old and New Ideas about the Human Sex Ratio." *Significance* 13.2(2016): 24–27.

Padula, Janice. "Proof and Rhetoric: The Structure and Origin of Proof." *Australian Senior Mathematics Journal* 30.1(2016): 45–54.

Parrott, Matthew. "Bayesian Models, Delusional Beliefs, and Epistemic Possibilities." *British Journal for the Philosophy of Science* 67(2016): 271–96.

Paseau, A. C. "Letter Games: A Metamathematical Taster." *The Mathematical Gazette* 100.549(2016): 442–49.

Pataut, Fabrice. "Comments on 'Parsimony and Inference to the Best Mathematical Explanation.'" *Synthese* 193(2016): 351–63. [see also Baker]

Paul, Annie Murphy. "The Coding Revolution." *Scientific American* 315.2(2016): 43–49.

Paulden, Tim. "Smashing the Racket." *Significance* 13.12(2016): 16–21.

Pisano, Raffaele. "Details on the Mathematical Interplay between Leonardo da Vinci and Luca Pacioli." *BSHM Bulletin—Journal of the British Society for the History of Mathematics* 31.2(2016): 104–11.

Pisano, Raffaele, and Paolo Bussotti. "A Newtonian Tale Details on Notes and Proofs in Geneva Edition of Newton's '*Principi*.'" *BSHM Bulletin—Journal of the British Society for the History of Mathematics* 31.3(2016): 160–78.

Politzer, Guy, and Jean Baratgin. "Deductive schemas with uncertain premises using qualitative probability expressions." *Thinking and Reasoning* 23(2016): 78–98.

Pollack, Henry. "Raking Leaves." *Consortium: Newsletter of the Center for Mathematics and Its Applications* 110(2016): 3–5.

Priest, Graham. "Logical Disputes and the A Priori." *Logique & Analyse* 236(2016): 347–66.

Rash, Agnes M., and Sandra Fillebrown. "Courses on the Beauty of Mathematics: Our Version of General Education Mathematics Courses." *PRIMUS* 26.9(2016): 824–36.

Rau, Pei-Luen Patrick, et al. "The Cognitive Process of Chinese Abacus Arithmetic." *International Journal of Science and Mathematics Education* 14(2016): 1499–516.

Rauff, James M. "The Algebra of Marriage: An Episode I Applied Group Theory." *BSHM Bulletin: Journal of the British Society for the History of Mathematics* 31.3(2016): 230–44.

Reinholz, Daniel L. "Developing Mathematical Practices through Reflection Cycles." *Mathematics Education Research Journal* 28(2016): 441–55.

Reys, Robert. "Some Thoughts on Doctoral Preparation in Mathematics Education." *Journal of Mathematics Education at Teachers College* 7.2(2016): 31–35.

Rice, Adrian, and Ezra Brow. "Commutativity and Collinearity: A Historical Case Study of the Interconnection of Mathematical Ideas." *BSHM Bulletin—Journal of the British Society for the History of Mathematics* 31.1(2016): 1–14, and 31.2(2016): 90–103.

Richeson, David. "Sugihara's Impossible Cylinder." *Math Horizons* 24.1(2016): 18–19.

Richezza, Victor J., and H. I. Vacher. "On a Desert Island with Unit Sticks, Continued Fractions and Lagrange." *Numeracy* 9.2(2016).

Ríordáin, Máire Ní, Jennifer Johnston, and Gráinne Walshe. "Making Mathematics and Science Integration Happen: Key Aspects of Practice." *International Journal of Mathematical Education in Science and Technology* 47.2(2016): 233–55.

Rochefort-Maranda, Guillaume. "How to Load Your Data Sets with Theories and Why We Do So Purposefully." *Studies in History and Philosophy of Science, A* 60(2016): 1–6.

Rochefort-Maranda, Guillaume. "On the Correct Interpretation of p Values and the Importance of Random Variables." *Synthese* 193(2016): 1777–93.

Roegel, Denis. "A Mechanical Calculator for Arithmetic Sequences (1844–1852)." *IEEE Annals of the History of Computing* 37.4(2015): 90–96 and 38.1(2016): 80–88.

Roitman, Pedro, and Hervé Le Ferrand. "The Strange Case of Paul Appell's Last Memoir on Monge's Problem." *Historia Mathematica* 43(2016): 288–309.

Rosa, Milton. "Humanizing Mathematics through Ethnomodelling." *Journal of Humanistic Mathematics* 6.2(2016): 3–22.

Roscoe, Matt B., and Joe Zephyrs. "Quilt Block Symmetries." *Mathematics Teaching in the Middle School* 22.1(2016): 18–27.

Ross, Drew M. "Determining the Use of Mathematical Geometry in the Ancient Greek Method of Design." *The Mathematical Intelligencer* 38.2(2016): 17–28.

Roth, Wolff-Michael. "On the Social Nature of Mathematical Reasoning." *For the Learning of Mathematics* 36.2(2016): 34–39.

Rowe, David E. "A Snapshot of Debates on Relativistic Cosmology, 1917–1924." *The Mathematical Intelligencer* 38.2(2016): 46–58.

Rowe, David E. "Otto Neugebauer and Richard Courant: On Exporting the Göttingen Approach to the History of Mathematics" *The Mathematical Intelligencer* 34.2(2012): 29–37.

Ruggieri, Eric. "Visualizing the Central Limit Theorem through Simulation." *PRIMUS* 26.3(2016): 229–40.

Ryan, Kathleen, and Brittany Shelton. "Statistics on the Bonus Round of *Wheel of Fortune*." *The College Mathematics Journal* 47.4(2016): 250–53.

Salmun, Haydee, and Frank Buonaluto. "The Catalyst Scholarship Program at Hunter College: A Partnership among Earth Science, Physics, Computer Science and Mathematics." *Journal of STEM Education* 17.2(2016): 42–50.

Saltelli, Andrea. "Young Statistician, You Shall Live in Adventurous Times." *Significance* 13.12(2016): 38–41.

Sarkar, Jyotimoy, and Mamunur Rashid. "Visualizing Mean, Median, Mean Deviation, and Standard Deviation of a Set of Numbers." *The American Statistician* 70.3(2016): 304–12.

Schappacher, Norbert, and Cordula Tollmien. "Emmy Noether, Hermann Weyl, and the Göttingen Academy." *Historia Mathematica* 43(2016): 194–97.

Schlimm, Dirk. "Metaphors for Mathematics from Pasch to Hilbert." *Philosophia Mathematica* 24.1(2016): 308–29.

Schubring, Gert. "Comments on a Paper on Alleged Misconceptions Regarding the History of Analysis: Who Has Misconceptions?" *Foundations of Science* 21(2016): 527–32.

Schwartz, Richard, and Sergei Tabachnikov. "Centers of Mass of Poncelet Polygons, 200 Years After." *The Mathematical Intelligencer* 38.2(2016): 29–34.

Schwarz, Alan. "ADHD: The Statistics of a 'National Disaster.'" *Significance* 13.6(2016): 20–23.

Scoth, Roberto. "Higher Education, Dissemination and Spread of the Mathematical Sciences in Sardinia (1729–1848)." *Historia Mathematica* 43(2016): 172–93.

Segura, Lorena, and Juan Matías Sepulcre. "A Rational Belief: The Method of Discovery in the Complex Variable." *Foundations of Science* 21(2016): 189–94.

Segura, Lorena, and Juan Matías Sepulcre. "Arithmetization and Rigor as Beliefs in the Development of Mathematics." *Foundations of Science* 21(2016): 207–14.

Shanno, Christine Ann. "Flipping the Analysis Classroom." *PRIMUS* 26.8(2016): 727–35.

Sharp, John. "Folding the Regular Pentagon." *BSHM Bulletin—Journal of the British Society for the History of Mathematics* 31.3(2016): 179–88.

Shields, Brit. "Mathematics, Peace, and the Cold War: Scientific Diplomacy and Richard Courant's Scientific Identity." *Historical Studies in the Natural Sciences* 46(2016): 556–91.

Shim, J. P., et al. "Phonetic Analytics Technology and Big Data: Real-World Cases." *Communications of the ACM* 59.2(2016): 84–90.

Sialaros, Michalis, and Jean Christianidis. "Situating the Debate on 'Geometrical Algebra' within the Framework of Premodern Algebra." *Science in Context* 29.2(2016): 129–50.

Siegmund-Schultze, Reinhard. "'Mathematics Knows No Races': A Political Speech That David Hilbert Planned for the ICM in Bologna in 1928." *The Mathematical Intelligencer* 38.1(2016): 56–66.

Silver, Daniel S. "Mathematical Induction and the Nature of British Miracles." *American Scientist* 104.5(2016): 296–303.

Simoson, Andrew. "Minimizing Utopia." *Math Horizons*, 23.3(2016): 18–21.

Skovsmose, Ole. "An Intentionality Interpretation of Meaning in Mathematics Education." *Educational Studies in Mathematics* 92.3(2016): 411–24.

Skovsmose, Ole. "What Could Critical Mathematics Education Mean for Different Groups of Students?" *For the Learning of Mathematics* 36.1(2016): 2–7.

Slater, Hartley. "Gödel's and Other Paradoxes." *Philosophical Investigations* 39.4(2016): 353–61.

Smadja, Ivahn. "On Two Conjectures That Shaped the Historiography of Indeterminate Analysis: Strachey and Chasles on Sanskrit Sources." *Historia Mathematica* 43(2016): 241–87.

Smith, Christopher E., and Paré, Jana N. "Exploring the Klein Bottle through Pottery." *Mathematics Teacher* 110.3(2016): 208–13.

Smith, Kim Bradford. "Counting People in a Conflict Zone." *Significance* 13.2(2016): 38–41.

Smith, Michael D. "Introducing Abelian Groups Using Bullseyes and Jenga." *PRIMUS* 26.3(2016): 197–205.

Somerville, Andrew. "A Bayesian Analysis of Peer Reviewing." *Significance* 13.2(2016): 32–36.

Staatsi, Juha. "On the 'Indispensable Explanatory Role' of Mathematics." *Mind* 125.500(2016): 1045–70.

Staecker, Christopher P. "Nice Neighbors: A Brief Adventure in Mathematical Gamification." *Math Horizons* 23.4(2016): 5–7.

Stanek, Edward J. "Predicting the Unknown—Sampling, Smoke, and Mirror." *Chance* 29.3(2016): 41–50.

Starr, Sonja. "Actuarial Risk Prediction and the Criminal Justice System." *Chance* 29.1(2016): 49–51.

Steinig, Rachel M. "Stop Ruining Math! Reasons and Remedies for the Maladies of Mathematics Education." *Journal of Humanistic Mathematics* 6.2(2016): 128–47.

Stojakovic, Vesna, and Bojan Tepavcevic. "Distortion Minimization: A Framework for the Design of Plane Geometric Anamorphosis." *Nexus Network Journal* 18(2016): 759–77.

Strickland, S., and B. Rand. "A Framework for Identifying and Classifying Undergraduate Student Proof Errors." *PRIMUS* 26.10(2016): 905–21.

Stylianides, Gabriel J., James Sandefur, and Anne Watson. "Conditions for Proving by Mathematical Induction to Be Explanatory." *The Journal of Mathematical Behavior* 43(2016): 20–34.

Sudakov, Ivan, et al. "Infographics and Mathematics: A Mechanism for Effective Learning in the Classroom." *PRIMUS* 26.2(2016): 158–67.

Sutil, Nicolas Salazar. "Mathematics in Motion: A Comparative Analysis of the Stage Works of Schlemmer and Kandinsky at the Bauhaus." *Dance Research* 32.1(2016): 23–42.

Suzuki, Jeff. "Confidence in the Census." *Math Horizons* 24.1(2016): 20–22.

Symeonidou, Ioanna. "Anamorphic Experiences in 3D Space: Shadows, Projections and Other Optical Illusions." *Nexus Network Journal* 18(2016): 779–97.

Taylor, Anne A., and Lucie M. Byrne-Davis. "Clinician Numeracy: The Development of an Assessment Measure for Doctors." *Numeracy* 9.1(2016).

Tezer, Murat, Meryem Cumhur, and Emine Hürse. "The Spatial-Temporal Reasoning States of Children Who Play a Musical Instrument, Regarding the Mathematics Lesson." *Eurasia Journal of Mathematics, Science & Technology Education* 12.6(2016): 1487–98.

Thunder, Kateri, and Robert Q. Berry III. "The Promise of Qualitative Metasynthesis for Mathematics Education." *Journal for Research in Mathematics Education* 47.4(2016): 318–37.

Triantafillou, Chrissavgi, Vasiliki Spiliotopoulou, and Despina Potari. "The Nature of Argumentation in School Mathematics and Physics Texts: The Case of Periodicity." *International Journal of Science and Mathematics Education* 14(2016): 681–99.

Trinchero, Roberto, and Giovanni Sala. "Chess Training and Mathematical Problem Solving: The Role of Teaching Heuristics in Transfer of Learning." *Eurasia Journal of Mathematics, Science & Technology Education* 12.3(2016): 655–68.

Tunstall, Samuel L. "Words Matter: Discourse and Numeracy." *Numeracy* 9.2(2016).

Tyner, Sam. "Using the R Package geomnet: Visualizing Trans-Atlantic Slave Trade of Africans, 1514–1866." *Chance* 29.3(2016): 4–16.

Usó-Doménech, José-Luis, Josué Antonio Nescolarde Selva, and Mónica Belmonte Requena. "Mathematical, Philosophical and Semantic Considerations on Infinity (I): General Concepts." *Foundations of Science* 21.4(2016): 615–30.

Vacher, H. I. "Grassroots Numeracy." *Numeracy* 9.2(2016).

Valfells, Sveinn. "Minting Money with Megawatts." *Proceedings of the IEEE* 104.9(2016): 1674–78.

Verburgt, Lukas M. "Duncan F. Gregory, William Walton and the Development of British Algebra: 'Algebraical Geometry,' 'Geometrical Algebra,' Abstraction." *Annals of Science* 73.1(2016): 40–67.

Verburgt, Lukas M. "Robert Leslie Ellis, William Whewell and Kant: The Role of Rev. H. F. C. Logan." *BSHM Bulletin—Journal of the British Society for the History of Mathematics* 31.1(2016): 47–51.

Vinsonhaler, Rebecca. "Teaching Calculus with Infinitesimals." *Journal of Humanistic Mathematics* 6.1(2016): 250–76.

Wade, Carol, et al. "A Comparison of Mathematics Teachers' and Professors' Views on Secondary Preparation for Tertiary Calculus." *Journal of Mathematics Education at Teachers College* 7.1(2016): 7–16.

Wagner, Roi. "Wronski's Foundations of Mathematics." *Science in Context* 29.3(2016): 241–71.

Wainer, Howard. "Defeating Deception." *Chance* 29.1(2016): 61–64.

Wainer, Howard. "Don't Try This at Home." *Significance* 13.1(2016): 22–23.

Wallace, Dorothy. "Teaching Quantitative Reasoning in an Exponential Decay Model." *Numeracy* 9.2(2016).

Wasserstein, Ronald L., and Nicole A. Lazar. "The ASA's Statement on *p*-Values: Context, Process, and Purpose." *The American Statistician* 70.2(2016): 129–33.

Weber, Keith. "Mathematical Humor: Jokes That Reveal How We Think about Mathematics and Why We Enjoy It." *The Mathematical Intelligencer* 38.4(2016): 56–61.

Weber, Keith, Timothy P. Fukawa-Connelly, Juan Pablo Mejía-Ramos, and Kristen Lew. "How to Help Students Understand Lectures in Advanced Mathematics." *Notices of the American Mathematical Society* 63.10(2016): 1190–93.

Weinberg, Aaron, Joshua Dresen, and Thomas Slater. "Students' Understanding of Algebraic Notation: A Semiotic Systems Perspective." *The Journal of Mathematical Behavior* 43(2016): 70–88.

Whitacre, Ian, et al. "Regular Numbers and Mathematical Worlds." *For the Learning of Mathematics* 36.2(2016): 20–25.

Wijeratne, Chanakya, and Rina Zazkis. "Exploring Conceptions of Infinity via Super-Tasks: A Case of Thomson's Lamp and Green Alien." *The Journal of Mathematical Behavior* 4(2016): 127–34.

Williams, Julian. "Alienation in Mathematics Education: Critique and Development of Neo-Vygotskian Perspectives." *Educational Studies in Mathematics* 92(2016): 59–73.

Wolchover, Natalie. "A Bird's-Eye View of Nature's Hidden Order." *Quanta Magazine* July 12, 2016, https://www.quantamagazine.org/20160712-hyperuniformity-found-in-birds-math-and-physics/.

Wu, Su-Chiao, and Fou-Lai Lin. "Inquiry-Based Mathematics Curriculum Design for Young Children: Teaching Experiment and Reflection." *Eurasia Journal of Mathematics, Science & Technology Education* 12.4(2016): 843–60.

Xia, Wen, et al. "A Comprehensive Study of the Past, Present, and Future of Data Deduplication." *Proceedings of the IEEE* 104.9(2016): 1691–710.

Ycart, Bernard. "Jakob Bielfeld (1717–1770) and the Diffusion of Statistical Concepts in Eighteenth Century Europe." *Historia Mathematica* 43(2016): 26–48.

Yeang, Chen-Pang. "Two Mathematical Approaches to Random Fluctuations." *Perspectives on Science* 24.1(2016): 45–72.

Yiwen, Zhu. "Different Cultures of Computation in Seventh Century China from the Viewpoint of Square Root Extraction." *Historia Mathematica* 43(2016): 3–25.

Yopp, David, A., and Rob Ely. "When Does an Argument Use a Generic Example?" *Educational Studies in Mathematics* 91(2016): 37–53.

Zazkis, Rina, and Igor' Kontorovich. "A Curious Case of Superscript (−1): Prospective Secondary Mathematics Teachers Explain." *The Journal of Mathematical Behavior* 43(2016): 98–110.

Zhao, Jiaying, and Yu Qi Yu. "Statistical Regularities Reduce Perceived Numerosity." *Cognition* 146(2016): 217–22.

Notable Book Reviews and Review Essays

Ackerberg-Hastings, Amy, reviews *Perfect Mechanics* by Richard Sorrenson. *Historia Mathematica* 43(2016): 104–5.

Alexanderson, Gerald L., reviews *Scientist, Scholar & Scoundrel* [Count Guglielmo Libri] by Jeremy M. Norman. *American Mathematical Monthly* 123.5(2016): 512–14.

Avigad, Jeremy, reviews *Logic's Lost Genius: The Life of Gerhard Gentzen* by Eckart Menzler-Trott and *Gentzen's Centenary* edited by Reinhard Kahle and Michael Rathjen. *Notices of the American Mathematical Society* 63.11(2016): 1288–92.

Bakker, Arthur, reviews *Networking of Theories as a Research Practice in Mathematics Education* edited by Angelika Bikner-Ahsbahs and Susanne Prediger. *Educational Studies in Mathematics* 93(2016): 265–73.

Best, Joel, reviews *The Math Myth, and Other STEM Delusions* by Andrew Hacker. *Numeracy* 9.2(2016).

Best, Joel, reviews *World War II in Numbers* by Peter Doyle. *Numeracy* 9.1(2016).

Bilgili, Alper, reviews *Science among the Ottomans* by Miri Shefer-Mossensohn. *Annals of Science* 73.4(2016): 449–50.

Bressoud, David M., reviews *The Math Myth, and Other STEM Delusions* by Andrew Hacker. *Notices of the American Mathematical Society* 63.10(2016): 1181–83.

Bullynch, Maarten, reviews *Histoire d'algorithmes* by Jean-Luc Chabert et al. *Historia Mathematica* 43(2016): 332–41.

Bullynch, Maarten, reviews *Newton and the Origin of Civilization* by Jed Z. Buchwald and Mordechai Feingold. *Historia Mathematica* 43(2016): 213–25.

Burgiel, Heidi, reviews *Creating Symmetry* by Frank A. Farris. *College Mathematics Journal* 47.3(2016): 228–31.

Carter, Jessica, reviews *Why Prove It Again?* by John W. Dawson Jr. *Philosophia Mathematica* 24(2016): 256–63.

Colyvan, Mark, reviews *The Mathematics of Love* by Hannah Fry. *Notices of the American Mathematical Society* 63.7(2016): 821–22.

Craig, William Lane, reviews *Natur und Zahl: Die Mathematisierbarkeit der Welt* by Bernulf Kanitscheider. *Philosophia Mathematica* 24(2016): 136–41.

Crampin, Stephanie, reviews four "recent mathematical histories." *BSHM Bulletin—Journal of the British Society for the History of Mathematics* 31.1(2016): 83–86.

Dawson, John W., Jr., reviews *David Hilbert's Lectures on the Foundations of Arithmetic and Logic, 1917–1933* edited by William Ewald and Wilfried Sieg. *Historia Mathematica* 43(2016): 105–7.

Ducheyne, Steffen, reviews *The Newton Papers* by Sarah Dry. *Historia Mathematica* 43(2016): 342–45.

Englebretsen, George, reviews *Articulating Medieval Logic* by Terence Parsons. *Ratio* 29(2016): 344–51.

Fey, James, reviews *The New Math* by Christopher J. Phillips. *Journal for Research in Mathematics Education* 47.4(2016): 420–22.

Folland, Gerald B., reviews *The Real and the Complex* by Jeremy Gray. *American Mathematical Monthly* 123.9(2016): 949–52.

Føllesdal, Dagfinn, reviews *After Gödel: Platonism and Rationalism in Mathematics and Logic* by Richard Tieszen. *Philosophia Mathematica* 24(2016): 405–24.

Garrity, Thomas, reviews *Differential Forms* by Steven H. Weintraub. *American Mathematical Monthly* 123.4(2016): 407–12.

Geerdink, L. M., and C. Dutilh Novaes review *Varieties of Logic* by S. Shapiro. *History and Philosophy of Logic* 37.2(2016): 194–96.

Gerovitch, Slava, reviews *Infinitesimal* by Amir Alexander. *Notices of the American Mathematical Society* 63.5(2016): 571–74.

Giansiracusa, Noah, reviews *Math on Trial* by Leila Schneps and Coralie Colmez. *Journal of Humanistic Mathematics* 6.2(2016): 207–24.

Glass, Darren B., reviews *The Fascinating World of Graph Theory* by Arthur Benjamin, Gary Chartrand, and Ping Zhan. *American Mathematical Monthly* 123.1(2016): 106–12.

Gray, Mary W., reviews *Math Goes to the Movies* by Burkard Polster and Marty Ross. *The Mathematical Intelligencer* 38.1(2016): 90–92.

Gray, Mary W., reviews seven books on Émilie du Châtelet by various authors. *The Mathematical Intelligencer* 38.4(2016): 70–77.

Greenwell, Raymond N., reviews *The Magic of Math* by Arthur Benjamin. *College Mathematics Journal* 47.4(2016): 307–11.

Growney, JoAnne, reviews *A New Index for Predicting Catastrophes* by Mahdur Anand. *Journal of Humanistic Mathematics* 6.2(2016): 200–6.

Heckman, Gert, reviews *Remembering Sofya Kovalevskaya* by Michèle Audin. *Nieuw Archief voor Wiskunde* 5/17.1(2016): 39–40.

Heeffer, Albrecht, reviews *The Emperor's New Mathematics* by Catherine Jami. *The Mathematical Intelligencer* 38.2(2016): 83–84.

Hogben, Leslie, and Mark Hunacek review *Linear Algebra Done Right* by Sheldon Axler. *American Mathematical Monthly* 123.6(2016): 621–24.

Hopkins, Brian, reviews *My Search for Ramanujan* by Ken Ono and Amir D. Aczel. *College Mathematics Journal* 47.5(2016): 475–80.

Hopkins, Burt, reviews *The Road Not Taken: On Husserl's Philosophy of Logic and Mathematics* by Claire Ortiz Hill and Jairo José da Silva. *Philosophia Mathematica* 24(2016): 263–75.

Humphreys, Paul, reviews *Reconstructing Reality* by Margaret Morrison. *Philosophy of Science* 83(2016): 627–33.

Jackson, Craig H., reviews *Mathematics and Climate* by Hans Kaper and Hans Engler. *American Mathematical Monthly* 123.3(2016): 304–8.

Jones, Ian, reviews *Algebra Teaching around the World* edited by Frederick K. S. Leung et al. *Educational Studies in Mathematics* 91(2016): 289–94.

Karaali, Gizem, reviews *Literacy & Mathematics* by Jay P. Abramson and Matthew A. Isom. *Numeracy* 9.2(2016).

Kidwell, Mark, reviews *Knots, Molecules, and the Universe* by Erica Flapan. *American Mathematical Monthly* 123.8(2016): 840–43.

Kobayashi, Mei, reviews *My Search for Ramanujan* by Ken Ono and Amir D. Aczel. *Notices of the American Mathematical Society* 63.8(2016): 890–92.

Koo, Alex, reviews *An Aristotelian Realist Philosophy of Mathematics* by James Franklin. *The Mathematical Intelligencer* 38.3(2016): 81–84.

Lamb, Evelyn, reviews *The Math Myth* by Andrew Hacker. *Slate Online* (March 29, 2016).

Lamprecht, Elizabeth A., reviews *Seduced by Logic* by Robyn Arianrhod. *Journal of Humanistic Mathematics* 6.1(2016): 277–84.

Lawrence, Snezana, reviews *Birth of a Theorem* by Cédric Villani. *BSHM Bulletin—Journal of the British Society for the History of Mathematics* 31.3(2016): 263–65.

Lawrence, Snezana, reviews *Handbook on the History of Mathematics Education* edited by Alexander Karp and Gert Schubring. *Educational Studies in Mathematics* 92(2016): 279–85.

Li, Yeping, and Jianxing Xu review *Student Voice in Mathematics Classrooms around the World* edited by Berinderjeet Kaur, Glenda Anthony, Minoru Ohtani, and David Clarke *Educational Studies in Mathematics* 91(2016): 141–48.

Mawhin, Jean, reviews *Hidden Harmony, Geometric Fantasies* by Umberto Bottazzini and Jeremy Gray. *La Matematica nella Società e nella Cultura* 8(2016): 157–68.

Mazur, Joseph, reviews *Genius at Play* [John H. Conway] by Siobhan Roberts. *The Mathematical Intelligencer* 38.1(2016): 78–81.

McCarty, Charles, reviews *Philosophy of Mathematics in the Twentieth Century* by Charles Parsons. *Philosophical Review* 125(2016): 298–302.

Mrozik, Dagmar, reviews *Pearls from a Lost City: The Lvov School of Mathematics* by Roman Duda. *Historia Mathematica* 43(2016): 207–15.

Nasifoglu, Yelda, reviews *Architecture and Mathematics from Antiquity to the Future*, two volumes edited by Kim Williams and Michael J. Ostwald. *BSHM Bulletin—Journal of the British Society for the History of Mathematics* 31.3(2016): 261–63.

Nunemacher, Jeffrey, reviews *Learning Modern Algebra* by Al Cuoco and Joseph J. Rotman. *American Mathematical Monthly* 123.2(2016): 205–8.

Panza, Marco, reviews *Infini, Logique, Géométrie* by P. Mancosu. *History and Philosophy of Logic* 37.4(2016): 396–99.

Pettigrew, Richard, reviews *Rigor and Structure* by John P. Burgess. *Philosophia Mathematica* 24.1(2016): 129–36.

Pincock, Christopher, reviews *Why Is There Philosophy of Mathematics at All?* by Ian Hacking. *British Journal for the Philosophy of Science* 67(2016): 907–12.

Presmeg, Norma, reviews *Semiotics as a Tool for Learning Mathematics* edited by Adalira Sáenz-Ludlow and Gert Kadunz. *Mathematical Thinking and Learning* 18.3(2016): 233–38.

Proulx, Jérôme, and Jean-François Maheux review *Approaches to Qualitative Research in Mathematics Education* edited by A. Bikner-Ahsbahs et al. *Mathematical Thinking and Learning* 18.2(2016): 142–50.

Rice, Adrian, reviews *Taming the Unknown* by Victor J Katz and Karen Hunger Parshall. *BSHM Bulletin—Journal of the British Society for the History of Mathematics* 31.1(2016): 81–83.

Richey, Matthew, reviews *An Introduction to Statistical Learning with Applications in R* by Gareth James et al. *American Mathematical Monthly* 123.7(2016): 731–36.

Ross, Peter, reviews *How Not to Be Wrong* by Jordan Ellenberg. *College Mathematics Journal* 47.2(2016): 146–52.

Schindler, Samuel, reviews *String Theory and the Scientific Method* by Richard Dawid. *Philosophy of Science* 83(2016): 453–58.

Sierpinska, Anna, reviews *Networking of Theories as a Research Practice in Mathematics Education* edited by Angelika Bikner-Ahsbahs and Susanne Prediger. *Mathematical Thinking and Learning* 18.1(2016): 69–76.

Sriraman, Bharath, reviews *Handbook of International Research in Mathematics Education* edited by Lyn D. English and David Kirshner. *Journal for Research in Mathematics Education* 47.5(2016): 552–56.

Stark, Philip B., reviews *Privacy, Big Data, and the Public Good* edited by Julia Lane et al. *American Statistician* 70.1(2016): 119.

Stern, David G., reviews *The Logical Must: Wittgenstein on Logic* by Penelope Maddy. *Analysis* 76.3(2016): 391–93.

Stillwell, John, reviews *Origins of Mathematical Words* by Anthony Lo Bello and *Enlightening Symbols* by Joseph Mazur. *Bulletin of the American Mathematical Society* 53.2(2016): 331–35.

Talbot, Richard, reviews *Manifold Mirrors: The Crossing Paths of the Arts and Mathematics* by Felipe Cucker. *Nexus Network Journal* 18(2016): 563–66.

Thomas, Emily, reviews *Descartes-Agonistes: Physico-mathematics, Method and Corpuscular-Mechanism 1618–33* by John Schuster. *Annals of Science* 73.1(2016): 112–14.

Trefethen, Lloyd N., reviews *The Princeton Companion to Applied Mathematics* edited by Nicholas J. Higham. *SIAM Review* 57.3(2015): 469–73.

Vinckier, Nigel, and Jean Paul Van Bendegem review *Strict Finitism and the Logic of Mathematical Applications* by Feng Ye. *Philosophia Mathematica* 24(2016): 247–80.

Wahl, Russell, reviews *The Palgrave Centenary Companion to* Principia Mathematica edited by Nicholas Griffin and Bernard Linsky. *History and Philosophy of Logic* 37.3(2016): 294–97.

Wragge-Morley, Alexander, reviews *Observing the World through Images* edited by Nicholas Jardine and Isla Fay. *Nuncius* 31.3(2016): 653–55.

Yap, Audrey, reviews *A Mathematical Prelude to the Philosophy of Mathematics* by Stephen Pollard. *Philosophia Mathematica* 24(2016): 275–77.

Zitarelli, David E., reviews *Mathematics across the Iron Curtain* by Christopher Hollings. *The Mathematical Intelligencer* 38.1(2016): 84–87.

Notable Interviews

Banerjee, Moulinath, and Bodhisattva Sen interview Michael Woodroofe. *Statistical Science* 31.3(2016): 433–41.

Barker, Richard, interviews G. A. F. Seber. *Statistical Science* 31.2(2016): 151–60.

Buckland, Stephen T., interviews Richard M. Cormack. *Statistical Science* 31.2(2016): 142–50.

Burton, Ernest DeWitt, interviews Stephen Stigler. *Significance* 13.2(2016): 42.

Delaigle, Aurore, and Matt Wand interview Peter Hall. *Statistical Science* 31.2(2016): 275–304.

Diaz-Lopez, Alexander, interviews Arlie Petters. *Notices of the American Mathematical Society* 63.4(2016): 376–77.

Diaz-Lopez, Alexander, interviews Colin Adams. *Notices of the American Mathematical Society* 63.10(2016): 1172–74.

Diaz-Lopez, Alexander, interviews Elisenda Grigsby. *Notices of the American Mathematical Society* 63.3(2016): 282–84.

Diaz-Lopez, Alexander, interviews Fernando Codá Marques. *Notices of the American Mathematical Society* 63.2(2016): 142–43.

Diaz-Lopez, Alexander, interviews Helen Moore. *Notices of the American Mathematical Society* 63.7(2016): 768–70.

Diaz-Lopez, Alexander, interviews Ian Agol. *Notices of the American Mathematical Society* 63.1(2016): 23–24.

Diaz-Lopez, Alexander, interviews Jordan Ellenberg. *Notices of the American Mathematical Society* 63.6(2016): 645–46.
Diaz-Lopez, Alexander, interviews Po-Shen Loh. *Notices of the American Mathematical Society* 63.8(2016): 905–8.
Diaz-Lopez, Alexander, interviews Sir Timothy Gowers. *Notices of the American Mathematical Society* 63.9(2016): 1026–28.
Kennedy, Stephen, interviews Helen G. Grundman. *Notices of the American Mathematical Society* 63.11(2016): 1258–62.
Koul, Hira L., and Roger Koenker interview Estate V. Khmaladze. *Statistical Science* 31.1(2016): 453–64.
Merow, Katherine, interviews Amie Wilkinson. *Math Horizons* 23.3(2016): 5–7.
Miller, Stephen D., interviews John Urschel. *Notices of the American Mathematical Society* 63.2(2016): 148–51.
Naus, Joseph, interviews Arthur Cohen. *Statistical Science* 31.3(2016): 442–52.
Persson, Ulf, interviews Jacob Murre. *Nieuw Archief voor Wiskunde* 5/17.1(2016): 58–62.
Persson, Ulf, interviews Luc Illusic. *Svenska Matematikersamfundet Medlemsutskicket*(2016): 10–34.
Pollatsek, Harriet, interviews David Goldberg. *Notices of the American Mathematical Society* 63.8(2016): 924–26.
Pollatsek, Harriet, interviews Karen Saxe. *Notices of the American Mathematical Society* 63.11(2016): 1298–99.
Raussen, Martin, and Christian Skau interview John F. Nash Jr. *Notices of the American Mathematical Society* 63.5(2016): 486–91.
Raussen, Martin, and Christian Skau interview Louis Nirenberg. *Notices of the American Mathematical Society* 63.2(2016): 135–40.
Reimann, Amy L., and David A. Reimann interview Anne Burns. *Mathematics Magazine* 89.5(2016): 375–77.
Reimann, Amy L., and David A. Reimann interview Bjarne Jespersen. *Mathematics Magazine* 89.1(2016): 55–57.
Reimann, Amy L., and David A. Reimann interview Dick Termes. *Mathematics Magazine* 89.4(2016): 290–92.
Reimann, Amy L., and David A. Reimann interview Robert Fathauer. *Mathematics Magazine* 89.2(2016): 220–22.
Tarran, Brian, interviews Cathy O'Neil. *Significance* 13.6(2016): 42–43.
Tarran, Brian, interviews Simon Tavaré. *Significance* 13.5(2016): 42–43.

Notable Lives in Mathematics: Profiles, Memorial Notes, and Obituaries

Borwein, Jonathan M. (1951–2016) *Educational Studies in Mathematics* 93(2016): 131–36.
Gerdes, Paulus (1952–2014) *Historia Mathematica* 43(2016): 129–32.
Gillispie, Charles Coulston (1918–2015) *Isis* 107.1(2016): 121–26.
Grattan-Guinnes, Ivor (1941–2014) *Isis* 107.4(2016): 875–79.
Guy, Richard K. (1916–) *The Mathematical Intelligencer* 38.4(2016): 19–22.
Hall, Peter Gavin (1951–2016) *Significance* 13.2(2016): 48.
Minsly, Marvin M. (1925–2016) *Nature* 530(2016): 282.
Steen, Lynn (1941–2015) *Numeracy* 9.1(2016).
Zeeman, Sir Erik Christopher (1925–2016) *Mathematical Gazette* 100.548(2016): 307–13.

Notable Journal Issues

"Integer Polynomials." *The American Mathematical Monthly* 123.4(2016).

"Association of Australian Mathematics Teachers 50th Anniversary." *Australian Mathematics Teacher* 72.3(2016).

"Forensic Statistics." *Chance* 29.1(2016).

"Ecology." *Chance* 29.2(2016).

"Communicational Perspectives on Learning and Teaching Mathematics." *Educational Studies in Mathematics* 91.3(2016).

"Mathematics Education and Contemporary Theory." *Educational Studies in Mathematics* 92.3(2016).

Historical Studies in the Natural Sciences 46.5(2016). [This is a special issue on mathematical superpowers after World War II.]

"Metacognition for Science and Mathematics Learning in Technology-Infused Learning Environments." *International Journal of Science and Mathematics Education* 14.2(2016).

"[Johannes von] Kries and Objective Probability." *Journal for General Philosophy of Science* 47.1(2016).

"The Nature and Experience of Mathematical Beauty." *Journal of Humanistic Mathematics* 6.1(2016).

"The Many Colors of Math: Engaging Students through Collaboration and Agency." *The Journal of Mathematical Behavior* 41(2016).

"Machine Learning and Music Generation." *Journal of Mathematics and Music* 10.2(2016).

"Mathematics Teachers as Partners in Task Design." *Journal of Mathematics Teacher Education* 19.2–3(2016).

"Interdisciplinary Contest in Modeling." *The Journal of Undergraduate Mathematics and Its Applications* 37.2(2016).

"[A New] Museum of the History of Sciences in Muscat, Oman." *Lettera Matematica* 4.2(2016).

"The Acquisition of Preschool Mathematical Abilities." *Mathematical Thinking and Learning* 17.2–3(2015).

"Mathematics Education and Mobile Technologies." *Mathematics Education Research Journal* 28.1(2016).

"Mathematical Knowledge for Teaching: Developing Measures and Measuring Development." *Mathematics Enthusiast* 13.1–2(2016).

"Teaching Mathematics Online." *Mathematics Teacher* 110.4(2016).

"Seed Concepts in Architecture and Mathematics." *Nexus Network Journal* 18.2(2016).

"Reconsidering Frege's Conception of Number." *Philosophia Mathematica* 24.1(2016).

"Mathematics and the Arts in the Mathematics Classroom." *PRIMUS* 26.4(2016).

"Teaching with Technology." *PRIMUS* 26.6(2016).

"Big Data [two parts]." *Proceedings of the IEEE* 104.1 and 11(2016).

"Governing Algorithms." *Science, Technology and Human Values* 41.1(2016).

"Sorting on arXiv." *Social Studies of Science* 46.4(2016).

"Models and Simulations." *Studies in History and Philosophy of Science* 55(2016).

"Indispensability and Explanation." *Synthese* 193.2(2016).

"Logic and the Foundations of Game and Decision Theory." *Synthese* 193.3(2016).

"Causation, Probability, and Truth." *Synthese* 193.4(2016).

"Learning with Data." *Technology, Knowledge and Learning* 21.1(2016).

"Cooperation and Competition in Sciences." *Zeitschrift für Geschichte der Wissenschaften, Technik und Medizin* 24.2(2016).

"Perception, Interpretation and Decision Making." *Zentralblatt für Didaktik der Mathematik* 48.1–2(2016).

"Cognitive Neuroscience and Mathematics Learning." *Zentralblatt für Didaktik der Mathematik* 48.3(2016).

"Improving Teaching, Developing Teachers and Teacher Developers, and Linking Theory and Practice through Lesson Study in Mathematics." *Zentralblatt für Didaktik der Mathematik* 48.4(2016).

"Survey on Research on Mathematics Education." *Zentralblatt für Didaktik der Mathematik* 48.5(2016).

"Mathematical Working Spaces in Schooling." *Zentralblatt für Didaktik der Mathematik* 48.6(2016).

"Methods for Helping Early Childhood Educators to Assess and Understand Young Children's Mathematical Minds." *Zentralblatt für Didaktik der Mathematik* 48.7(2016).

Acknowledgments

As always, I thank first the authors whose contributions are included in this anthology and the original publishers of the pieces.

At Princeton University Press, Vickie Kearn guided me with care and invaluable assistance throughout the phases of the book; Lauren Bucca solved the copyright issues; Nathan Carr oversaw the production process; and Paula Bérard copyedited the manuscript. Thank you to all.

After failing to find a mathematics teaching job, I reoriented myself professionally; I went back to school. I am discovering the dynamic and exciting academic field of librarianship and information. To sustain myself during these studies, I continued to support undergraduate mathematics courses, this time courtesy of the generous opportunities offered by the mathematics department at Syracuse University. Many thanks to Graham Leuschke, Uday Banerjee, Moira McDermott, Julie O'Connor, Kim Ann Canino, and Sandra Ware.

In my new field of studies, I have worked on many issues and projects. It would have been difficult to transition without the passionate professionals who taught me during the past year, including Barbara Stripling (to whom I dedicate this book), Jill Hurst-Wahl, and Caroline Haythornthwaite at Syracuse University's iSchool and from Camille Andrews at the Cornell University Mann Library.

During the work for this book, my family and I lived through tumultuous times, some of dramatic dimensions, impossible to characterize even succinctly here. I am grateful to the people who steered the situation from what looked like a train wreck toward slightly more manageable circumstances.

Credits

"Mathematical Products" by Philip J. Davis. Originally published in *Mathematics, Substance, and Surmise*, edited by P. J. Davis and E. Davis. Heidelberg, Germany: Springer International, 2016, pp. 69–74.

"The Largest Known Prime Number" by Evelyn Lamb. Originally published in *Slate*, January 22, 2016. Copyright © 2016 The Slate Group LLC

"A Unified Theory of Randomness" by Kevin Hartnett. Originally published in *Quanta Magazine*, August 2, 2016, an editorially independent publication of the Simons Foundation.

"An 'Infinitely Rich' Mathematician Turns 100" by Siobhan Roberts. Originally published in *Nautilus*, September 30, 2016. Reprinted by permission of NautilusThink.

"Inverse Yogiisms" by Lloyd N. Trefethen. Originally published in *Notices of the American Mathematical Society* 63.11(2016): 1281–85. Copyright © 2016 American Mathematical Society. Reprinted by permission of the American Mathematical Society.

"Ramanujan in Bronze" by Gerald L. Alexanderson. Originally published in *Bulletin of the American Mathematical Society*, Volume 53, Number 4, October 2016, pages 673–679. Copyright © 2016 American Mathematical Society. Reprinted by permission of the American Mathematical Society.

"Creating Symmetric Fractals" by Larry Riddle. Originally published in *Math Horizons*, Vol. 24.2(2016): 18–21. Published by permission of Math Horizons. Copyright © 2016 Mathematical Association of America. All Rights Reserved.

"Projective Geometry in the Moon Tilt Illusion" by Marc Frantz. Reprinted with kind permission from Springer Science+Business Media: *The Mathematical Intelligencer* 34.6(2016): 28–34. Copyright © 2016 Springer Science+Business Media New York. DOI 10.1007/s00283-016-9661-2

"Girih for Domes: Analysis of Three Iranian Domes" by Mohammadhossein Kasraei, Yahya Nourian, and Mohammadjavad Mahdavinejad. Reprinted with kind permission from Springer Science+Business Media: *Nexus Network Journal* 18(2016): 311–321. Copyright © 2016 Springer Science+Business Media New York. DOI 10.1007/s00004-015-0282-4

"Why Kids Should Use Their Fingers in Math Class" by Jo Boaler and Lang Chen. Originally published in *The Atlantic Online*, April 13, 2016. Reprinted by permission of *The Atlantic*.

"Threshold Concepts and Undergraduate Mathematics Teaching" by Sinéad Breen and Ann O'Shea. Originally published in *PRIMUS: Problems, Resources, and Issues in Mathematics Undergraduate Studies*, 26.9(2016): 837–47, reprinted by permission of Taylor & Francis Ltd, http://www.tandf.co.uk/journals

"Rising above a Cause-and-Effect Stance in Mathematics Education Research" by John Mason. Reprinted with kind permission from Springer Science+Business Media: *Journal of Mathematics Teacher Education* 19(2016): 297–300. Copyright © 2016 Springer Science+Business Media New York DOI 10.1007/s10857-016-9351-1

"How to Find the Logarithm of Any Number Using Nothing but a Piece of String" by Viktor Blåsjö. Originally published in *The College Mathematics Journal* 47.2(2016): 95–100. Copyright © 2016 Mathematical Association of America. All Rights Reserved.

"Rendering Pacioli's Rhombicuboctahedron" by Carlo H. Séquin and Raymond Shiau. Originally published in *Journal of Mathematics and the Arts* 9.3–4(2016): 103–110, reprinted by permission of Taylor & Francis Ltd, http://www.tandf.co.uk/journals

"Who Would Have Won the Fields Medal 150 Years Ago?" by Jeremy Gray. Originally published *Notices of the American Mathematical Society* 63.3(2016): 269–74. Copyright © 2016 American Mathematical Society. Reprinted by permission of the American Mathematical Society.

"Paradoxes, Contradictions, and the Limits of Science" by Noson S. Yanofsky. Originally published in *American Scientist* 104.3(2016): 166–73. Reprinted by permission of Sigma Xi, The Scientific Research Society.

"Stairway to Heaven: The Abstract Method and Levels of Abstraction in Mathematics" by Jean-Pierre Marquis. Reprinted with kind permission from Springer Science+Business Media: *The Mathematical Intelligencer* 38.3(2016): 41–51. Copyright © 2016 Springer Science+Business Media New York DOI 10.1007/s00283-016-9672-z

"Are Our Brains Bayesian?" by Robert Bain. Originally published in *Significance* 13.4(2016): 14–19. Reprinted by permission of John Wiley & Sons.

"Great Expectations: The Past, Present, and Future of Prediction" by Graham Southorn. Originally published in *Significance* (13.2(2016): 15–19) the magazine of the Royal Statistical Society and the American Statistical Association, www.significancemagazine.com. Reprinted by permission of John Wiley & Sons on behalf of the Royal Statistical Society and the American Statistical Association.